Lakeland

Paul Gannon

 Pesda Press LTD

www.pesdapress.com

Cover Photo: Langdale,
© iStockphoto.com Finn Brandt

First published in Great Britain 2009 by Pesda Press
Unit 22, Galeri
Doc Victoria
Caernarfon
Gwynedd
LL55 1SQ

© Copyright 2009 Paul Gannon

ISBN: 978-1-906095-15-4

Printed in Poland, produced by Polska Book.

Reg, Debbie, Roland, Barbara;
Charlie, Antony.
Lakeland lovers all.

Classic Lakeland scenery – a glacial valley carved into tough volcanic rock. Upper Langdale from Rosset Pike.

Contents

Introduction

The Lake District is England's best-loved mountain landscape. It has long been celebrated by poets and writers, adored by fellwalkers and tourists, and venerated by connoisseurs of scenic natural landscapes. The raw compelling beauty of its long narrow lakes, steep-sided valleys and soaring fells has drawn a multitude of visitors over the last three centuries or so. Many of the twelve million people who today visit the area each year do so in order to trek up and down the steep fellsides. They follow in the footsteps of many earlier visitors as well as many generations of inhabitants.

Some of those many fellwanderers wonder why the landscape looks like it does. What created this spectacular tension between fell and valley, rough craggy ridge and smoothly sloping fellside, flat valley floor and hummocky low land, ribbon-shaped valley lake and tiny, rounded mountain tarn?

Photo 0.1 Scafell Pike and Scafell, seen from Esk Pike.

This book aims to help fellwalkers and other visitors answer these questions. The book is in two parts. The first records the fascinating story of how colliding continents, violent volcanoes, irresistible mountain-building forces and intensive glaciation combined to shape the landscape we see nowadays. I attempt to explain why these volcanoes occurred, what sort of rocks they created and how to recognise signs of mountain-building and glaciation on the fells and in the valleys. The second part of the book contains fifteen recommended walks of differing levels of difficulty, with a wide variety of geological features, allowing you to enjoy consistently excellent views of the best of Lakeland's wonderful natural scenery.

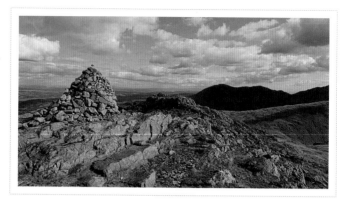

Photo 0.2 | Pike o' Blisco. Lakeland's popular mountain scenery is the creation of fire and ice.

Great pleasure can be had from walking the fells by appreciating something of the combined influence of volcanoes, glaciation and erosion on the present-day landscape. The book concentrates on what you can see as you walk, without recourse to a hand lens or other special tools. The aim is to enable the reader to identify major and minor features in Lakeland and elsewhere. The landscape is the product of many different forces and factors. Identifying and unravelling these is an endlessly fascinating pastime for the landscape lover.

While geology is fascinating, it relies on a complex scientific terminology. My aim has been to minimise the use of jargon to aid the hillwalker's understanding. This has meant some simplification of the jargon and a pruning of the detail to a minimum. However, I'm afraid we will still have to deal with a minimal set of terms such as 'plate tectonics' and 'fissure-vents', 'magma' and 'intrusive' rocks and 'ice sheets' and 'moraines'.

Geological jargon, when first used, is set in single quotes (e.g. 'lava'). Although many of these terms are defined in the Glossary, I have also sometimes assumed that the meaning is often obvious from the context. In a few cases single quotes have also been used to identify everyday idiom, such as a 'blob' of magma.

While reading the book, the reader may well find it useful to have a large scale map of the Lake District (such as the OS Touring Map 3 1:63,360 or the Harvey 1:40,000) to hand. Both these maps convey a good impression of the topography and will assist in locating the places mentioned in the text (with the help of the index of place names at the back of the book which includes grid references).

CHAPTER 1

Sedimentary Scenery

A good place to start our look at the landscapes of Lakeland is atop Castle Head, a small prominence located a short distance from the bustling town of Keswick. From here it is possible to look over an area that has taken some 500 million years to be shaped into its present-day scenery.

Looking to the north towards the Skiddaw massif and to the north-west to the hills around Causey Pike, the fells are made up of 'sedimentary' rocks. They were created nearly half a billion years ago from material eroded from other rocks, carried away by streams and rivers, then laid down in the sea and later compressed and hardened into solid rock. These sedimentary rocks today form softly shaped fells and ridges and offer superb walking conditions.

Photo 1.1

Sedimentary rocks on the right on Catbells give way to volcanic rocks to the centre and left. Derwentwater, a glacial lake, in the foreground. See also Photo 2.3.

Looking from our vantage point on Castle Head towards the south beyond Derwentwater and to the south-east at Walla Crag and High Rigg, the landscape is more rugged, frequently with knobbly fell tops and plenty of rough crags. These are volcanic or 'igneous' rocks, erupted in various forms during major Earth movements ('tectonic activity') about 460 million years

Photo 1.2 | View from Catbells towards the Skiddaw massif in the distance. These hills are made up of sedimentary rocks created some 500 million years ago, formed from grains of sand and silt eroded from higher ground and deposited in the sea, eventually hardening into rock.

ago. The deposits erupted from the volcanoes were later compressed and hardened into volcanic rock. Castle Head itself is made up of such volcanic rock. Indeed, as we shall see in Chapter 2, geologists believe it is probably the solidified 'plug' of an ancient volcanic 'vent' (see also Walk 11). These hills too offer excellent walking, but often with more ups and downs and a greater variety of rock outcrops to be negotiated.

About 300 million years ago all these rocks, sedimentary and volcanic, were compressed and pushed upwards to form a massive mountain range. This range was at least as high as the Alps, possibly even higher. During this time, some of the rocks were also transformed into 'slate' by the immense pressures of mountain-building.

In the millions of years since then, the hills have been eroded by water and ice. The wide flat lowland before you, some of it occupied by the stunningly beautiful Derwentwater, is a product of very recent erosion by glacier ice. The 'ice age', during which the lake was created, only relented its frozen grip on the area about 10,000 years ago, a mere moment in geological terms. Glaciers also carved deep lake-filled valleys and other features such as 'corries' and 'arêtes' high up in the mountains, creating many of the most popular and haunting features of the Lakeland landscape such as Blea Water and Striding Edge.

Since the end of the ice age the area has become ever more intensely inhabited, with Keswick and the network of roads the most conspicuous evidence of the human contribution to the landscape from this vantage point. Less evident from here is the effect on the landscape of the extraction of slate for building and especially roofing material.

Standing here on this ancient volcanic plug, we can therefore see a considerable variety of geological features and influences on the landscape, ranging from about 500 million years ago to today. A lot has happened to the restless Earth in that half a billion years time-span, most spectacularly through the movement of 'tectonic plates'.

We live on the surface of the Earth's outermost layer, the crust, which can be up to 45km deep under the continents and 10km deep under the oceans. Immediately below the

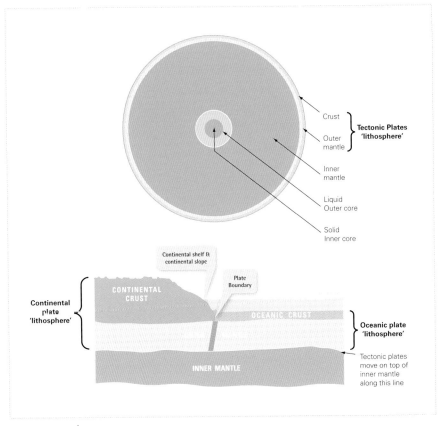

Diagram 1.1 | The Earth's core, mantle and crust (not to scale)

crust is the outer layer of the Earth's 'mantle'. The outer layer of the mantle combines with the crust to form the 'lithosphere' (literally the 'sphere of rocks'). The lithosphere, however, is not a coherent united layer all the way around the Earth. Instead, it is split into a series of interlocking 'tectonic plates'. These plates effectively float on the inner mantle and can move around the globe changing position very, very slowly at between just 0.2cm and 5cm a year.

The force that provides the power to move the tectonic plates is created by heat emanating from the Earth's innermost part, its molten 'core'. The heat causes convection currents in the solid mantle, with hot, pressurised solid rock actually 'flowing' from the centre rising towards the Earth's surface, cooling and then sinking. Although rock appears to us as absolutely solid, at very great pressures and high temperatures it can actually behave like a fluid and flow. Hot rock near the core rises, while cooler rock near the top falls back down again.

The tectonic plates (and the continental and oceanic crusts that they carry) sit on top of this inner mantle and are driven around by the convection currents, sometimes towards or away from each other, sometimes alongside one another and at times into collision with each other. As we will see, it was the collision of tectonic plates that created Lakeland's volcanic heartland.

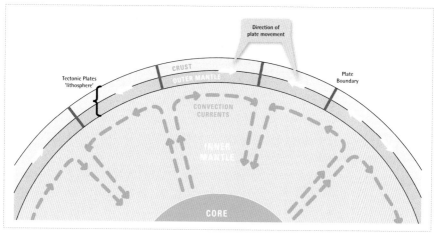

Diagram 1.2 | Convection currents and the movement of tectonic plates. Heat from the core generates convection currents in the mantle. Although the rocks of the mantle are solid, they are under such great pressure and high temperature that the rock flows, transferring heat from the core towards the surface and driving the tectonic plates (the upper mantle and the crust). The plate motions are responsible for volcanic activity on plate margins. No differentiation is shown for oceanic crust (thinner and more dense) and continental crust (thicker and less dense).

Diagram 1.3 | Avalonia, Laurentia and the Iapetus Ocean.

The Earth is thought to be about 4,500 million years old. For about the last four billion years, the Earth's surface has undergone a constant shifting of the plates with all the continents sometimes converging into one massive continent and at other times splitting up into several smaller continents. We take up the story about 500 million years ago, still an incredibly long time, but only just over a tenth of the Earth's overall history.

Half a billion years ago, the chunk of continental crust that would one day become Lakeland was then actually located somewhere south of the equator. It moved slowly and fitfully towards its present position as part of the overall pattern of movement of the tectonic plates driven by the convection currents. However, we need not worry about this aspect of the story. We concentrate instead on the relative movements of oceanic and continental plates with respect to each other, for this was what led directly to the creation of volcanic rocks and mountain-building.

The area of our interest sat at the western edge of one of the continents. This landmass is given the name of Avalonia by geologists (but it could also be seen as proto-north-west Europe). To the west lay an ocean named the Iapetus Ocean (proto-Atlantic ocean) and beyond it lay another continent, Laurentia (proto-north America).

It is important to appreciate that the continents often extend quite some way under sea with a gently sloping 'continental shelf' and, at its limit, a steep 'continental slope', which runs down to a deep trench where the underlying continental plate actually meets the plate

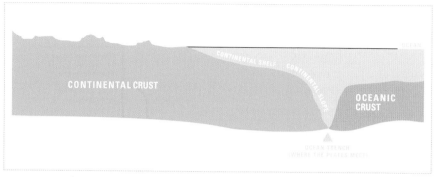

Diagram 1.4 | Continental shelf and continental slope.

carrying the oceanic crust. Most of the geological action in Lakeland took place in this continental shelf zone, sometimes undersea, sometimes above it.

It is from those seas, lapping onto the shore above the continental shelf on the edge of Avalonia, that we can begin to find rocks cropping out today on the surface in parts of Lakeland, creating a distinctive scenery. For tens of millions of years rocky material, eroded from high ground above sea level on the continent, was transported by rivers to the seas on the continental shelf and then deposited there as 'sediment'. Each year, say, a millimetre of material would be eroded from above sea, carried downhill by streams and rivers and then dumped below sea level. This may not seem to be very much at all, hardly noticeable over the run of a human lifetime. Yet if the processes of erosion and sedimentation continue for a very long time, then a very considerable thickness of sedimentary material can be laid down (or very considerable amounts of high land on the continent can be eroded away). After much time, these sediments are dried, compressed and hardened into 'sedimentary rock' (in a process called 'lithification').

One millimetre in a year equals 1cm after 10 years, 1m after 1,000 years, 100m after 100,000 years and 1,000m after 1 million years. Even allowing for compression and drying after being deposited, this means that over long-term geological periods of time extremely thick layers of sedimentary rock can be laid down. Of course, equally thick layers of rock from above ground can be eroded away, reducing high mountains to lowly stumps.

The sedimentary rocks deposited in the Lakeland seas are called the 'Skiddaw Group' (previously known as the 'Skiddaw Slates') and, all told, are several thousand metres thick. They must, therefore, have been deposited over many millions of years with endless amounts of material being eroded and dumped, year in, year out, on a scale that defeats our capacity to comprehend.

The Skiddaw Group comprises a wide variety of different types of sedimentary rocks: sandstone, siltstone and mudstone. The definition of the type of sedimentary rock is determined by the size of the eroded particles: large particles form sandstone, smaller particles form siltstone and yet even smaller particles form mudstone. The definition therefore tells us little about the original nature of the eroded rocks and only about their 'grain' size.

The rocks of the Skiddaw Group are mainly from the early part of what geologists call the Ordovician era, which extends from about 490 million years ago to about 430 million years ago. This is the time frame in which both the sedimentary and volcanic rocks of Lakeland were laid down.

It's useful to know these names as they are often mentioned in geology books, but we need not worry too greatly about them as they are not really significant to our story. The eras are determined by the types of fossils that occur in the rocks and not by the types of rock. Each new era represents a major change in the fossil record. The Cambrian era, which preceded the Ordovician, saw an explosion in fossil types as the variety of life forms increased dramatically in number and complexity. New types of fossils became common in the Ordovician

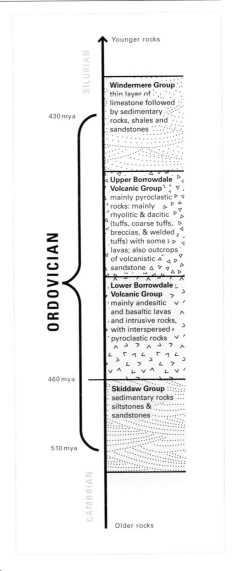

Diagram 1.5 | Geological 'time column' for Central Lakeland (not to scale). mya (millions of years ago).

era, such as the well known 'trilobite'. However, due to the extent of tectonic and volcanic activity in the Lakeland area, very few fossils have survived. Fossils are therefore not part of our story.

The seas into which the sediments were dumped were not always quiet backwaters. Periodically, they were affected by tectonic activity in the form of earthquakes. When earthquakes occurred, transmitting powerful tremors through the Earth, the still soft or partly dried sediments (which had been deposited on the sloping continental shelf) were easily dislodged, causing them to slump or even to slide down the sloping seafloor. As this happened repeatedly, it led to the mixing of the different rock types. The result of all this combined sedimentation and tectonic activity was to create a wide variety of sedimentary rock types and to muddle them up in an intensely complex manner. It has therefore taken geologists many decades to unravel the evidence.

Before looking at the rocks created by these aeons of sedimentation, it would be useful to make a general comment on the problem of recognising different rock types. This is not an easy task. Geologists use specialised equipment to determine rock types and have developed a highly complex (and all but impenetrable) language to describe rocks. The hillwalker, however, must make do with what they can see out on the fells.

One important complicating factor is 'weathering' (chemical changes caused by exposure of rock to the atmosphere). In general, weathering tends to cause all rocks to become greyish in colour. Another confounding factor is the presence of 'lichen' (a combination of fungi and algae) which soon start to form a coloured coat on the rocks, hiding the original surface and making identification difficult. On occasion, if you come across a rock that has recently been broken or smashed, you can usually see a quite different looking internal colour and perhaps structure. However, as you walk around, nearly all of what you see is weathered, lichen-encrusted rock. I will therefore concentrate on pointing out features that can help with identifying weathered rock, without the need for geochemical analysis or the use of magnifying lens.

Features which we will look at, such as 'bedding' and 'folding' in sedimentary rocks, mean it is possible to identify some rock types. (There are other features that we will look at in the next two chapters for 'igneous' or volcanic rocks.)

The Skiddaw Group rocks make up the fells of the Skiddaw massif and also of the north-western fells (Catbells, Causey Pike, Grasmoor, Hopegill Head and Grisedale Pike among them) as layer after layer of sedimentation was deposited and slowly transformed into hardened rock. Each layer is known in geological jargon as a 'bed'. Beds are usually parallel layers of

Photo 1.3

Bedding in sedimentary rocks. The top layer of each individual bed records an event, such as drying out under tidal or seasonal conditions or perhaps an earthquake.

horizontal bands, often of varying thickness. However, geology is endlessly complex and there are exceptions to this generalisation, as we shall see. Beds can also be tilted from the horizontal and even folded by later tectonic movements.

Bedding is a classic feature of sedimentary rocks, but is not always present. This is because some interruption to the sedimentation cycle is needed to allow the uppermost surface of a bed to dry out and become etched into the later rock. Some signs of bedding are visible in some rocks within the Skiddaw Group (see Walk 1), but not everywhere (see Photo 1.3).

It is also possible to see rocks where the bedding has become disturbed and has 'slumped' due to earthquakes or slides of rock down the sloping continental shelf. Mountain-building processes can sometimes lead to folding and bending of the beds and the creation of another useful feature, 'cleavage'. Cleavage is often visible in fine-grained sedimentary rocks (such as siltstone and mudstone) and sometimes looks like bedding, but in fact it has a quite different origin being created by 'mountain-building' forces (see chapter 4). Cleavage is sometimes known as 'slaty cleavage' and is indeed the defining characteristic of slate. Photo 1.4 shows bedding, folding, 'faulting' (breaks in the rock) and cleavage in a single small rock outcrop; the outcrop is partly weathered and partly covered by grey lichen (this outcrop can be seen on Walk 1 when descending from Whiteside).

Diagram 1.6 + Photo 1.4 | This sedimentary rock (Whiteside, about 0.75m high) displays bedding (running from upper left to lower right), folding of the bedding (best seen upper centre right) and cleavage (lines running almost vertically upper left to lower right); there are also some cracks or 'faults' running upper left to lower right.

The Skiddaw Group rocks were also affected by the intrusion of molten rock ('magma') which rose from the underlying mantle and collected underneath the sedimentary rocks. This occurred towards the end of the major volcanic episodes and during the subsequent mountain building episode (Chapters 2, 3 and 4). These pulses of hot molten magma did not make their way to the surface, but remained intruded between beds of rocks below the surface. There they cooled and solidified very slowly indeed, remaining hot enough for long enough to 'cook' the surrounding rocks, toughening them considerably and enabling them to resist erosion better than non-cooked sedimentary rocks.

It is this combination of repeated laying down of beds of deposits, uplift, compression, folding and tilting during mountain-building and hardening from slow-cooling underlying igneous intrusions, followed by millions of years of erosion by ice and water, that has produced the fells of the Skiddaw massif and the north-west Lakes area. Sedimentary rocks are often fairly weak compared with volcanic rocks and are easily eroded away by the power of glaciers in frozen periods. However, having been toughened by slow cooling of hot magma, the Skiddaw Group rocks have been able to resist the forces of erosion.

Indeed, Skiddaw is one of five fells in Lakeland which is higher than 900m, and the only one made up of sedimentary rock (the others – Scafell Pike, Scafell, Bowfell and Helvellyn – are all made from volcanic rock). The Skiddaw massif, which includes the popular peak of Blencathra, is a raised plateau-like area of ground. The area is incised by mainly gently sloping valleys, except for the north-south trending Glenderaterra valley which is somewhat steeper sided (due to glacial action). The slopes of the hills of the Skiddaw massif are curvaceous and sinuous, except on the southern rim where steeper slopes face the glacial plain now occupied in part by Keswick and surrounding lowland. The summit area of Skiddaw is only slightly curved, creating a wide and long summit plateau that undulates gently.

Photo 1.5

The gently curved summit plateau of Skiddaw.

The north-western fells, however, are rather different. They generally form fairly level ridges, narrow and long, with fairly constant slopes on either side and with some high points along the ridge. The ridges are definitely not knife-edge narrow, but are much sharper than Skiddaw's fairly flat summit ridge. As a result, they make for supremely exhilarating but pretty safe walking such as along the ridges between Causey Pike and Eel Crag, between Grisedale Pike and Sand Hill, between Knott Rigg and Ard Crags, between Hopegill Head and Whiteside (see Walk 1) or, for a gentle introduction to the delights of sedimentary ridge-walking, between Maiden Moor and Catbells (see Walk 11).

It is often possible to see rough rock faces cut across bedding (or cleavage) exposed on one side of the ridges, and the smooth rock faces on the other where the rock has been

Photo 1.6 | Sinuous curves characterise the landscape of the bulk of the Skiddaw massif.

Photo 1.7 | Looking north from Hopegill Head along a classic sedimentary ridge of the north-western fells to Ladyside Pike, centre, and Swinside, centre.

Photo 1.8 | Looking towards the Causey Pike ridge from Catbells,
another classic ridge in the north-western fells.

eroded away by ice and water on a weakness formed by a tilted bedding or cleavage plane. This effect is seen in an inverted form in some deep valleys such as Gasgale Gill on Walk 1.

Photo 1.9 | Bedding which has been cut
across on one side of the Catbells ridge.

Photo 1.10 | The same rock as in Photo 1.9,
but from the other side of the Catbells ridge
where bedding is parallel to the slope.

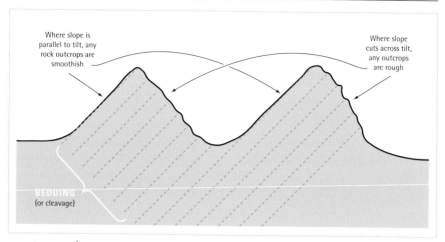

Diagram 1.7 | How the slope angle relative to the bedding affects the appearance of outcrops.

We have seen how, for a period as long as 50 million years, layer upon layer of sediment eroded from high ground on the continent was laid down on the seafloor. Movement of tectonic plates sparked earthquakes which caused the semi-solidified sediments to slip and slump, mixing them up. Somewhat later, massive forces during mountain-building episodes folded and twisted the rocks, turning some of them into slate. We also saw how the injection of hot magma underneath the rocks baked them, toughening them up so that they could resist the forces of erosion.

Ineluctably, those forces of erosion still found weaknesses to exploit. These were often weaknesses in the rock's own stucture and thus we have, after 500 million years, the land-scapes of this part of Lakeland with the sinuous curves of the Skiddaw massif and the fine ridges of the north-western fells. These are restrained but rewarding hills which repay the effort of traipsing over them, with superb walking conditions and magnificent views. While on these ridges one's eye is inevitably drawn by the views to the south and the much more extrovert and jagged tops of Lakeland's volcanic rocks. These were suddenly erupted into and onto the sedimentary scenery of Lakeland during the relatively short period of just a couple of million years.

CHAPTER 2

Lava Landscapes

We saw in the previous chapter that our area was sited on a continental shelf of Avalonia, jutting out under sea level into the Iapetus Ocean. We also read of tectonic activity which caused earthquakes sending tremors through the Earth, rudely shaking the land and dislodging some of the sediments that had been laid down on the sloping seabed. In this chapter, we look at why all this violent activity occurred and at the landscape created by these dramatic events.

Up until about 500 million years ago the Iapetus Ocean was getting wider as the continents on its margin, Avalonia to the east and Laurentia to the west, were driven away from each other by convection currents within the mantle. The oceanic plate between the continents expanded to fill the gap as they moved away. This was possible as there was a massive ridge in the middle of the ocean. Fissures in the oceanic plate at this ridge allowed vast quantities of molten lava to pour out onto the undersea surface through 'fissure-vents'. This is how new oceanic rock and oceanic crust are created.

As long as Avalonia and Laurentia continued to move apart from one another this did not matter very much, as the ocean could simply grow larger and larger. Meanwhile, the aeons of sedimentation continued on the shores of Avalonia.

Photo 2.1
Lava landscapes,
Grange Fell

About 500 million years ago things changed. The convection currents in the mantle began to drive Avalonia and Laurentia towards, instead of away from, each other. There followed a very long period, of about 100 million years in all, when the continental plates pushed in from either side. While this was happening at the margins, things went on unchanged in the middle of the ocean. Just as much lava poured out of the fissure-vents, pushing newly formed oceanic plate out against the continental plates that were now beginning to push inwards.

The plates probably moved at just a few centimetres per year, so progress was extremely slow yet relentless. When two plates meet they continue to be driven by their underlying convection currents; something must give.

The oceanic plate, which is much thinner and denser than continental plate, is usually pushed underneath the thicker and less dense continental plate. This is known as 'subduction' and is the cause of earthquakes and volcanic activity. In some circumstances, it can lead to the creation of volcanic mountain ranges such as the Andes and Cascades. In other instances, it may lead to the development of an 'arc' of island volcanoes on the edge of the continental plate, such as the Indonesian island chain. It was such an island arc that was created when Avalonia and Laurentia started to move towards each other, squeezing in the ocean between them.

The convection currents push relentlessly so that the two plates grind and scrape against one another, while friction prevents them moving continuously. Instead, sudden jerky move-

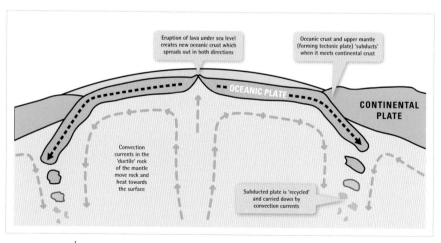

Diagram 2.1 | Creation of new oceanic crust and plate.

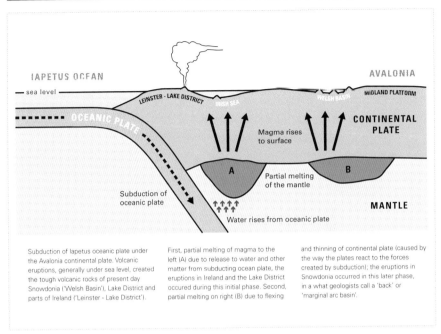

IAPETUS OCEAN

AVALONIA

— sea level —

LEINSTER - LAKE DISTRICT

IRISH SEA

WELSH BASIN

MIDLAND PLATFORM

OCEANIC PLATE

CONTINENTAL PLATE

Magma rises to surface

A

B

Partial melting of the mantle

Subduction of oceanic plate

MANTLE

Water rises from oceanic plate

Subduction of Iapetus oceanic plate under the Avalonia continental plate. Volcanic eruptions, generally under sea level, created the tough volcanic rocks of present day Snowdonia ('Welsh Basin'), Lake District and parts of Ireland ('Leinster - Lake District').

First, partial melting of magma to the left (A) due to release to water and other matter from subducting ocean plate, the eruptions in Ireland and the Lake District occured during this initial phase. Second, partial melting on right (B) due to flexing

and thinning of continental plate (caused by the way the plates react to the forces created by subduction); the eruptions in Snowdonia occurred in this later phase, in a what geologists call a 'back' or 'marginal arc basin'.

Diagram 2.2 | Subduction of Iapetus oceanic plate under the Avalonia continental plate.

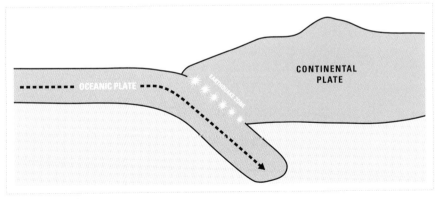

CONTINENTAL PLATE

OCEANIC PLATE

EARTHQUAKE ZONE

Diagram 2.3 | Earthquakes generated by subduction, as the oceanic plate is slowly pushed against and under the continental plate. The plates move toward one another at perhaps 2cm per year. Although the oceanic plate is pushed continuously from behind, friction at the meeting point of the plates prevents continous movement there. The plates lock together until the pressure behind is too great and the plates move suddenly, perhaps 2 metres every 100 years, sparking off earthquakes.

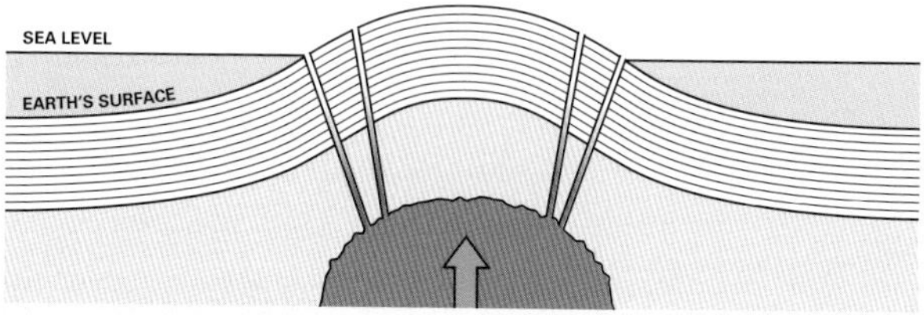

STAGE 1

1 Magma begins to rise, forcing itself into existing rock formations.

2 Extra bulk of magma accommodated by 'doming' of uppermost layers of rock.

3 Magma starts to rise up old fault lines towards surface.

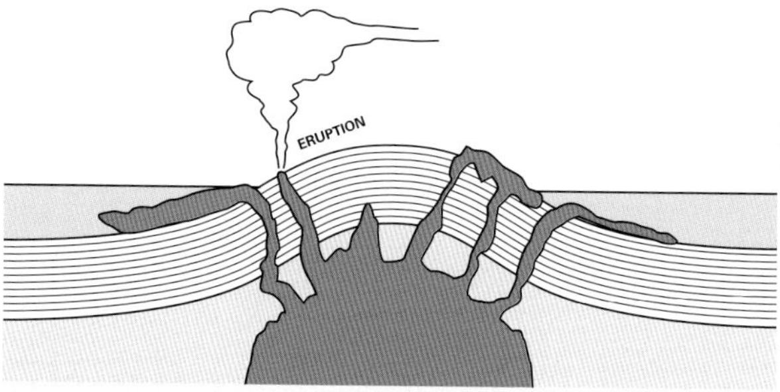

SEA LEVEL

EARTH'S SURFACE

STAGE 2

1 Magma reaches surface as pyroclastic eruptions (example excludes lava eruptions)

2 Above sea level, eruption of pyroclastic matter results in 'air-fall tuffs', or 'ash-surge tuffs' or 'ash-flow tuffs' (& breccias)

3 Below sea level, eruption of pyroclastic matter results in 'ash-surge tuffs' or 'ash-flow tuffs' (and breccias)

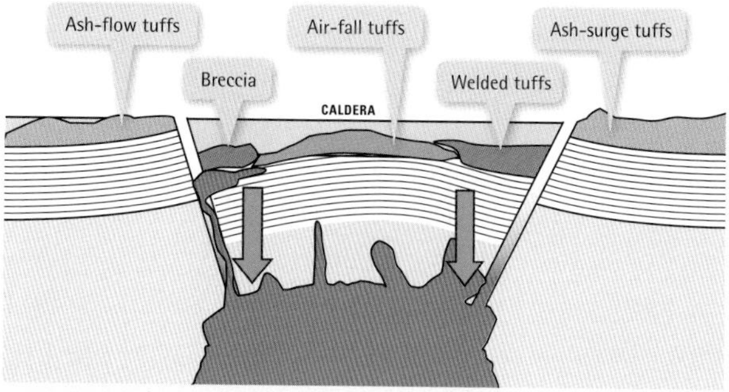

ERUPTION

STAGE 3

1 The magma is substantially erupted, leaving a void below, which is filled by collapse along the old fault lines and fissure-vents, leaving a 'caldera'.

2 Various eruptive matter becomes volcanic tuffs and breccias, laid down in and around caldera.

3 In a final stage, some of remaining magma will rise up fault lines into new volcanic rock formations as 'intrusive rhyolite'

Ash-flow tuffs

Breccia

CALDERA

Air-fall tuffs

Welded tuffs

Ash-surge tuffs

Diagram 2.4 | Rising magma causes 'doming' and eventual 'caldera' collapse.

ments occur when the pressures build up and can no longer be resisted; the oceanic plate suddenly shifts downwards a few metres or so, accommodating the pressures built up by say a few thousand years of plate movement. This transmits tectonic energy through the crust in the form of an earthquake. Earthquakes can set off tsunamis ('tidal waves') and slumping of deposits.

Even more dramatic than earthquakes, subduction also generates volcanic activity. As the oceanic plate descends under the continental plate it plunges down into the underlying mantle, dragging sea water down with it. For a while, the subducting plate retains its coherence even as it plunges into the mantle, creating a wedge of plate protuding some way down into the mantle (where it is eventually 'recycled'). The sea water then starts to escape from the descending plate surface, working its way into the overlying wedge of mantle. The water has the effect of reducing the temperature at which the mantle rocks become molten. So subduction results in the creation of a massive underground 'blob' of molten magma (or possibly repeated generations of blobs of molten magma over time).

The liquid molten magma is less dense than the surrounding solid mantle so it begins to rise through any weaknesses formed by 'faults' or cracks in the Earth. The magma rises towards the overlying continental plate, eventually forming a chamber of magma just below the surface rocks. These surface rocks may be pushed upwards into a dome shape if the magma is unable to find its way to the surface easily. This created the arc of islands through which Lakeland's volcanoes vented their fury.

This is the basic story of what happened in the Lake District area some 460 million years ago. Subduction of the Iapetus oceanic plate under the Avalonian continental plate led to the creation of a massive blob (or perhaps a series of blobs over time) of molten magma under the surface rocks, eventually forcing its way up to erupt at the surface.

The volcanic episode in Lakeland was part of a wider cycle of subduction that created other volcanic rocks in areas of Wales (Snowdonia, the Berwyns, Cader Idris, near Builth Wells and western Pembrokeshire), southern Ireland (Leinster) and parts of England (Caer Caradoc and also in boreholes in the midlands and east). This whole cycle probably lasted as long as 50 million years, but each area was only affected for a relatively short time within that overall period. The Lakeland volcanic episode occurred roughly in the middle of the overall cycle, lasting fewer than 10 million years, at around about 460 million years ago. The volcanic episode that created Snowdonia occurred about 30 million years later and lasted perhaps 3–5 million years.

The episode of volcanic activity in Lakleand occurred in two major sub-cycles, each consisting of a number of smaller cycles, and each of these with several different centres of eruption. Each centre of eruption probably consisted of several 'vents' and each vent probably experienced a number of individual eruptive events. Each eruptive event probably consisted of a group of connected eruptions (and probably accompanied by earthquakes, blastwaves and mudflows). Eruptive events might occur fairly continuously for a few thousand years, or last a few days and be repeated only once every few thousand years. An eruptive centre might have a lifetime of a few tens or hundreds of thousand years, perhaps even a million years.

Photo 2.2 | Lava flow (Hawaii). Photo © iStockphoto.com.

The two Lakeland volcanic sub-cycles created what geologists for a long time called the 'Lower Borrowdale Volcanic Group' and the 'Upper Borrowdale Volcanic Group'. Though named after Borrowdale, these rocks are not confined to the Borrowdale area but form much of the central core of the fells of Lakeland, including Great Gable, Scafell Pike, Bowfell, the Langdale Pikes, Haystacks, Fairfield, Helvellyn, High Street and Old Man of Coniston, to name only the best-known peaks.

Photo 2.3 | A classic 'cone' volcanoe, Andes. The first Lakeland eruptions were vented from low profile 'shield' volcanoes. Photo Jim Young.

The Borrowdale Volcanic Group contains a vast variety of rock types. We will look at the Lower BVG eruptions and the rocks they produced in this chapter and at the Upper BVG in the following chapter. One point to note is that some geologists now tend to call the Lower BVG rocks the Birker Fell formation, as there are some classic outcrops on Birker Fell. However, I will stick with the familiar Lower BVG name (also because the Birker Fell formation tag is used inconsistently).

The Lower BVG were the first volcanic rocks in the area and are mainly (but not exclusively) the product of extensive flows of lava.

Volcanic eruptions at the surface of the Earth basically fall into two different types: 'effusive' and 'pyroclastic'. Effusive eruptions pour out molten lavas, familiar nowadays to viewers of the BBC or Discovery Channel. Orange-red magma cascades onto the surface from some primeval breach in the Earth's surface and then, as lava, flows gracefully downhill covering potentially vast areas. This is, by and large, how the Lower BVG was formed.

On the other hand, pyroclastic eruptions involve the sudden and explosive release of gases and fragments within the magma. On reaching the surface the molten magma explodes into vast numbers of fragments, ranging in size from tiny 'ash' to massive 'bombs', which are spewed out either upwards or to the side. Great clouds of material are created, which then settle on the ground or under sea and harden into volcanic rocks over time. There are some Lower BVG rocks that were produced by pyroclastic eruptions, but they make up a small part of the whole group. On the other hand, the Upper BVG (see Chapter 3) are mainly (but not entirely) pyroclastic rocks.

Lavas and pyroclastic rocks (such as 'tuffs') are both 'extrusive' igneous rocks. However, not all the magma reaches the surface. Some remains below the surface, intruded between other rocks, where it slowly cools to form 'intrusive' igneous rock (as we saw in Chapter 1 where some magma cooled having been injected into the crust below the sedimentary rocks of the Skiddaw Group). Some parts of the Lower BVG may be rocks formed by magma which did not rise all the way to the surface.

An arc of Lower BVG rocks lies south of the Skiddaw Group sedimentary rocks, from around Ullswater in the east to Wastwater in the south-west and south to Harter Fell.

The classic concept of the volcano is of a cone; this is in fact only one type of volcano. In the Lake District there was not one large cone volcano but a series of lowish volcanoes, and the eruptions took place from a variety of different types of vents. There was probably one vent every 4km or so across the whole area and probably also beyond. The lava was erupted from low-lying 'shield' volcanoes which poured out masses of flowing lava forming thick sheets. The gentle slopes allowed the lava to flow for up to several kilometres, smothering everything beneath them, including previous sheets of lava which, over time, cooled and hardened into solid rock.

These sheets are anything between 10m and 300m thick, but generally are within the range of 250–300m. They flowed for anything from a few hundred metres up to 5km or more. All told, the depth of the combined lava sheets in Lakeland is as much as 2.8km at its thickest.

There are several potential sites of the vents, or of the intrusive rocks now exposed in the 'plumbing conduits' (a network of faults) leading to them, such as Castle Head near Keswick and in Eskdale, Lingcove, Cockley Beck and the upper Duddon valley, but most are now untraceable.

In some places it is quite easy to see the lava sheets stacked on top of one another, forming a characteristically 'step' or 'trap' (from the German word for step) shape. Examples include High Rigg (Walk 12), Border End (Hard Knott), Birker Fell (between Crook Crags and Green Crags) and in the area of Scoat Tarn. The sheets are separated by thin beds of pyroclastic or sedimentary rock which have been eroded into benches, while the lavas stand proud as steps. This type of feature is known as 'trap topography'.

Photo 2.4 (right) | Castle Head: a volcanic plug now exposed at the surface (foreground) and standing proud of the surrounding landscape.

Photo 2.5 (below) | Trap topography exposed on the side of High Rigg. The sheets of lava are clearly visible and form steps on the slope and skyline.

The trap topography displayed on the eastern side of High Rigg, seen from St John's in the Vale, is obvious (see Photos 2.5 and w12.10). However, not all lava outcrops result in the well-defined trap topography seen on High Rigg. The feature only really develops if the sheets of lava rock have been tilted at a fairly gentle angle, allowing the bench or tread to be etched out. There are plenty of places where the rocks are Lower BVG lavas but where you cannot see a clear example of trap topography. It is, all the same, possible to see lava sheets in some of the fellsides. Looking from Haystacks to the flanks of Fleetwith Pike or from High Rigg to Clough Head, for example, it is possible to see the lava sheets within the distant fellsides.

The Lower BVG rocks are known as 'basaltic' and 'andesitic' lavas, generally darkish in colour (see Chapter 3).

We could say that trap topography is a 'macro-feature' of lava, best seen from some distance and lost as a clear feature when you get close enough to see the rock in detail. There are only a few clearly distinguishing 'micro-features' in the individual outcrops when seen close up. All the same some do exist and can be identified fairly easily by the hillwalker. One of the most enthralling feature is called 'flow-banding'. This is the product of lava flows and is at its most impressive when the flows have slightly solidified and then, for some reason, been moved again. This can produce quite magnificently folded lines in the rocks. There are some spectacular flow-banding lines to be seen on and around the summit area of Haystacks (Walk 2) and Border End (see Photos 2.6, 2.7, w2.3 and w2.4). On Haystacks, the bands of lava have been folded into concertina shapes.

Photo 2.6 (top left) | Flow-banding in lava flow – Border End

Photo 2.7 (top right) | Flow-banding in lava flow – Border End

Another feature, created when slightly solidified lavas are given new impetus to move, is known as 'breccia' and the process of its creation as 'brecciation'. Lumps of rock are created when the lava moves and the more solid layers, near the upper and lower edges, break up into bits. Sometimes the original flow banding is visible in the resulting lumps of broken lava.

A 'breccia' is in fact any type of rock which consists of lumps of other rocks larger than 64mm in size. Breccias have different origins, and are not just derived from 'brecciated' lava. Indeed, breccia is also a very common type of pyroclastic rock as we will see in the next chapter. One way that can help you to decide whether a breccia is lava or pyroclastic is to

Photo 2.8 | Breccia.

check if the lumps stand proud of the surrounding rock (pyroclastic) or if they are slightly lower than the surface of the surrounding rock (lava). Such rules must be handled carefully as 'brecciated lava' can appear in various different forms.

Sometimes lavas have lots of small holes in them ('vesicles', thus 'vesicular lava') where gas has escaped from the molten lava once it has erupted onto the surface. This is a good indicator that the rock is a lava.

Flow-banding, breccia and vesicles are all features to look for in the Lower BVG rocks. There will often be few features to help distinguish those rocks, such as the exposed rocks on Castle Head which are quite plain.

It is important to note that not all Lower BVG rocks are lavas although the great bulk of them certainly are. The repeated lava eruptions were punctuated by generally small and sporadic pyroclastic explosions, which produced a variety of rock types. We will look in more detail at pyroclastic rocks in the next chapter.

The Lower BVG rocks often produce fells with broad summit plateaus, although the plateau surface is usually highly irregular with innumerable small ups and downs, often with a scattering of small tarns held in the intervening hollows between small rises and outcrops.

Photo 2.9 | Lava landscape on the summit plateau of High Rigg.

Photo 2.10 | The Jaws of Borrowdale – gateway to the volcanic rocks of modern day Lakeland.

This scenery is exemplified by the ground crossed in Walks 2 (Fleetwith Pike and Haystacks) and 12 (High Rigg) with summit areas made up of Lower BVG rocks. A few pyroclastic outcrops can also be seen, especially on Fleetwith Pike, as well as thick lava sheets.

The whole line of Lower BVG fells presents a sharp craggy face, a 'scarp' slope, to the north and west of the overall outcrop. This is best seen on the eastern side of Derwentwater and to its south, with the crags running from Walla Crag to the 'Jaws of Borrowdale'. This is one of Lakeland's most loved views.

In the early days of tourism to the Lake District, visitors were often fearful of setting foot beyond the Jaws of Borrowdale, worried that the terrain would be impassably rough and that they would most probably perish. Today fellwalkers seek out the jagged landscapes for the pleasure they bring. Although the summits of the Lower BVG rocks are slightly more restrained than those of the Upper BVG, they can sometimes provide excellent fellwalking conditions and wonderful views. That said, at other times they can be tiresome and massive boggy morasses.

Indeed, the summits of the Lower BVG rocks are not as high as those of the later, Upper BVG rocks. Only a few of the Lower BVG peaks are numbered among the best known of Lakeland's fells. Nonetheless, there are peaks such as Haystacks which are immensely impressive from some angles if rather inconspicuous from others. Haystacks was the favourite fell of the legendary writer of guidebooks on Lakeland, Wainwright. It is commonly claimed that his ashes were scattered, at his request, in the area of the Innominate Tarn on Haystacks.

Photo 2.11 | Lava landscape: view east from Walla Crag towards Castlerigg Fell and Bleaberry Fell.

Haystacks is one of those mountains that benefits from being slightly lower than many of the surrounding fells, offering dramatic views of the higher points all around. It is quite an easy mountain to climb if you start from the top of Honister Pass, but this hides the steep, craggy mountain slopes over Scarth Gap and on the flanks. Approached instead after a climb up the north-western ridge of Fleetwith Pike or from Scarth Gap you are left in no doubt that, while comparatively small, these are still truly rough mountains (Walk 2).

Around the summit area you can find some particularly striking lavas which have been folded like a concertina. This is a feature known as flow-banding and in this case has been disturbed and crumpled into these striking shapes (see Photos w2.3 and w2.4). Something happened to the lava when it was still soft and before it had fully hardened into rock, causing it to bend and fold. Outcrops of rock formed from lava can be seen all over the tilted summit plateau (not all of it showing flow-banding, however). Haystacks is a riot of minor crags and requires a certain amount of hand-steadying when crossing up and down the innumerable rocky dips and rises. It is a classic lava landscape.

The contrast between landscapes is easily seen looking north-west from the flanks of Grey Knotts with Haystacks and the High Stile ridge beyond (see Photo 2.11). The Lower BVG

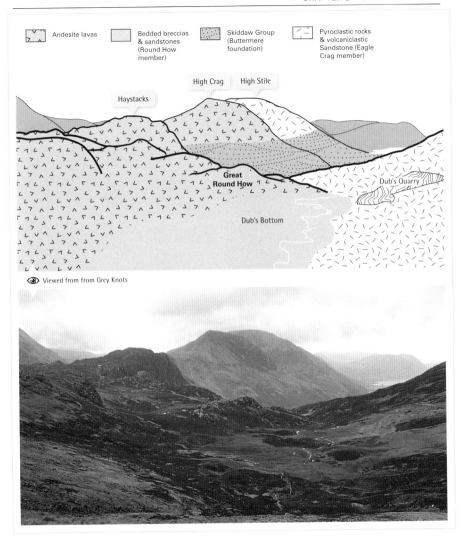

Andesite lavas

Bedded breccias & sandstones (Round How member)

Skiddaw Group (Buttermere foundation)

Pyroclastic rocks & volcaniclastic Sandstone (Eagle Crag member)

High Crag   High Stile

Haystacks

Great Round How

Dub's Quarry

Dub's Bottom

Viewed from from Grey Knots

Diagram 2.5 + Photo 2.12 | Lava landscape: from the summit area of Haystacks (upper left) to the centre left of the photograph; pyroclastic and sedimentary rocks (centre and right, including Dub's slate quarry); and pyroclastic landscape of High Crag and High Stile (centre distance).

rocks extend beyond Haystacks and Scarth Gap up to and past the summit of High Stile to that of Red Pike, but these outcrops are mainly pyroclastic rather than lava. The lava has produced the knobbly landscape seen on Haystacks and elsewhere, while the pyroclastic rocks have resulted in the pointed ridge seen in the distance.

These pyroclastic rocks were formed fairly early during the period of the Lower BVG and are known as the Eagle Crag member (see diagram 2.5). This is made up of different types of sedimentary rocks (siltstone and sandstone) as well as pyroclastic rocks. Some of these sedimentary rocks have been transformed into slate (see Chapter 4) and form the slate beds that are exploited at Honister and just below Haystacks itself in the heart of Lakeland. The recently re-opened Dub's Quarry is visible in the photograph and occurs in a different outcrop from that exposed on High Stile.

The cycle of volcanic activity that produced the Lower BVG rocks eventually drew to a close and a second cycle began, producing the Upper BVG (mainly pyrolastic) rocks. In the next chapter we will look at the Upper BVG volcanic activity and how to spot these rocks in the field today.

CHAPTER 3

Tuff Terrain

Once the Lower Borrowdale Volcanic Group sub-cycle of eruptions drew to a close, a new sub-cycle started which was to last for about five to ten million years with several episodes of volcanicity. However, only a few of these new eruptions produced more flows of lava. Instead, this second sub-cycle was predominantly explosive 'pyroclastic' eruptions. These spewed out a variety of volcanic material and generated localised earth movements, creating the mix of pyroclastic and sedimentary rocks that today make up the central, eastern and southern fells of Lakeland (including the highest peaks such as Scafell Pike, Bowfell, Helvellyn, High Street and Coniston Old Man).

The scenery produced by the rocks created during this new sub-cycle is quite different from the lava landscapes we looked at in the previous chapter. Compare, for example, the

Photo 3.1 | Tuff Terrain – view to the Scafell massif from Scoat Fell.

topography seen in Photos 2.1 and 3.1. The reason for this is that the magma which erupted in this new sub-cycle, which produced the rocks known as the Upper Borrowdale Volcanic Group (Upper BVG), had a different chemical make-up from that of the Lower BVG (see chapter 2).

As we saw, the Lower BVG rocks are generally (but not always) darkish in colour and are known as 'basalts' or 'andesites'. They are rocks with a relatively low proportion of silica in their chemical make-up. Rocks with a higher proportion of silica are known as 'dacites' and 'rhyolites'. The Upper BVG rocks are generally (but not wholly) rhyolitic or dacitic.

It is worth devoting a few sentences to what lies behind these differences, but this does mean dealing with a few technical terms. I will keep the number of terms to a minimum practical level, but as there are over 1,500 names for different types of volcanic rock, some simplification is necessary.

Magma type:	Rhyolitic	Dacitic	Andesitic	Basaltic
Grain size:				
Fine	rhyolite*	dacite*	andesite*	basalt*
Medium	micro-granite	micro-granodiorite	diorite	dolerite
Coarse	granite*	granodiorite	trachy-andesite	gabbro

Diagram 3.1 | Igneous magma and rock types. *mentioned in this book.

The key point for our purposes is that basaltic and andesitic magma usually flows fairly easily, which means it has low 'viscosity'. Dacitic and rhyolitic magma, on the other hand, is much more viscous and does not flow easily.

What this means is that basaltic and andesitic magmas tend to rise easily through any plumbing conduits in the Earth's crust all the way to the surface, then flow steadily outwards in fairly continuous rivers of lava forming the often thick but regular sheets we looked at in the last chapter.

However, the viscous dacitic and rhyolitic magmas are more reluctant to flow, tending to congeal and solidify in the vents as they rise up towards the surface. This prevents the escape of the enormous pressures that are built up as the underlying mass of molten magma continues to move upwards. Instead there is a concentration of pressure underneath

Diagram 3.2 |
Relationship
between
viscosity and
proportion of
silica among
different
magma types.

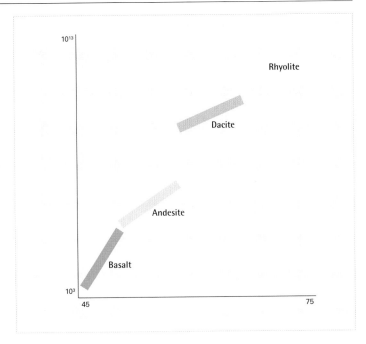

the semi-solidified plug formed by the congealed magma. When enough pressure has been bottled up it may become too great for the congealed magma to hold back any longer and the gases within the magma escape explosively – like opening a well-shaken bottle of fizzy water or champagne.

The magma changes from being a molten mass containing dissolved gases to being a gaseous mass containing fragments of molten magma (and maybe also surrounding rock ripped away in the explosion). This erupts onto the Earth's surface, either under the sea or into the atmosphere in great steaming clouds of various sorts. These fragments of various sizes settle to the ground or seafloor and become the material that hardens into pyroclastic rocks of various sorts.

It is, I'm afraid, necessary to add that this does not mean that basalts and andesites are always lavas, nor that rhyolites are always pyroclasts. In geology there are always exceptions to such generalisations and in Lakeland you can indeed come across basaltic and andesitic pyroclastic rocks as well as some dacitic and even comparatively rare rhyolitic lavas. More often than not, however, you will encounter lavas that are basaltic and andesitic and pyroclastic rocks that are dacitic and rhyolitic.

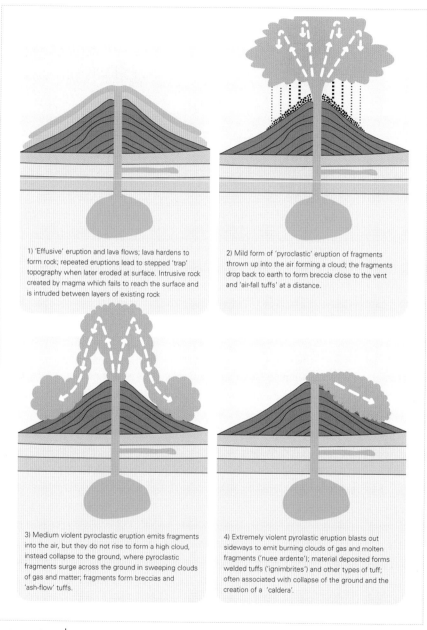

1) 'Effusive' eruption and lava flows; lava hardens to form rock; repeated eruptions lead to stepped 'trap' topography when later eroded at surface. Intrusive rock created by magma which fails to reach the surface and is intruded between layers of existing rock

2) Mild form of 'pyroclastic' eruption of fragments thrown up into the air forming a cloud; the fragments drop back to earth to form breccia close to the vent and 'air-fall tuffs' at a distance.

3) Medium violent pyroclastic eruption emits fragments into the air, but they do not rise to form a high cloud, instead collapse to the ground, where pyroclastic fragments surge across the ground in sweeping clouds of gas and matter; fragments form breccias and 'ash-flow' tuffs.

4) Extremely violent pyroclastic eruption blasts out sideways to emit burning clouds of gas and molten fragments ('nuee ardente'); material deposited forms welded tuffs ('ignimbrites') and other types of tuff; often associated with collapse of the ground and the creation of a 'caldera'.

Diagram 3.3 | Lava, pyroclastic rocks and intrusive rocks.

Pyroclastic eruptions can throw up enormous clouds of tiny particles of 'ash' that slowly fall through the air, blanketing the land surface. Alternatively, an eruption of hot gas and fragments can explode upwards until it loses its momentum, then collapse back to Earth and surge sideways smothering the surrounds. Another type of eruption can blast out sideways, emitting glowing clouds of gas and molten material that flow at hundreds of kilometres an hour, instantaneously incinerating everything in their path.

These different types of pyroclastic eruption – air-fall, surge, and flow – produce a wide variety of rock types which we can generically call 'tuff'. We will look at some of the pyroclastic rock types which can be spotted on the fells in more detail later in this chapter.

The latest view among geologists is to divide the Upper BVG into various phases, such as the Scafell Caldera succession, the Duddon Basin succession and the Rydal area succession. Each succession represents a linked series of eruptions that formed a cycle of 'caldera creation'. The Scafell Caldera was the first such cycle and illustrates the different stages common to caldera-creating cycles.

One point to note is that 'an eruption' may consist of a set of closely spaced events including one or more emissions of erupted material. When Mount St. Helens in the north-west of the USA erupted in 1980, the first activity was a series of earthquakes on the 20th March, followed by a series of small eruptions over the next two months. At the same time, the northern flank of the mountain was lifted by about 150 metres as magma rose up inside the mountain. On the 18th May, a large earthquake severely shook the entire upper mountain and led to the collpase of three massive chunks of the northern slope, accompanied by a series of pyroclastic eruptions all on the same day. The northern half of the cone at the summit of the mountain was blasted away by several eruptions on that day leaving a crater 600 metres deep as millions of cubic metres of material were expelled along with lumps of the pre-existing mountain. A blast wave flattened forest in an area of 600 cubic kilometres, a column of gas and ash rose to 16 kilometres (10 miles), pyroclastic flows dumped hot matter on surrounding land, mudflows buried vast areas and new lakes were created. Many more eruptions, including some lava flows, followed in the months after the cataclysmic eruption in May 1980. All of this can be treated as a single 'eruption'. Looking back, geologists have identified at least 60 layers of material left by previous eruptions over the last few thousand years.

In the Scafell Caldera succession a series of large pyroclastic eruptions took place, probably from fissure-vents to the north-west and/or west of the central fells. These eruptions produced andesitic and dacitic tuffs.

Another series of eruptions followed shortly after but were much bigger, releasing perhaps as much as 150 million cubic metres of pyroclastic material. A lot of the rocks produced were of a type known as 'welded tuff', which requires extremely hot and gaseous conditions in the eruption. As we will see later, welded tuffs are one of the easier types of volcanic rocks

Middle Dodd Dacite formation Esk Pike formation	Rydal succession	
Lincomb Tarns formation Seathwaite Fells formation	Duddon Basin succession	**UPPER BORROWDALE VOLCANIC GROUP**
Lingmell formation Airy's Bridge formation Whorneyside formation	Scafell caldera succession	
Birker Fell formation Pyroclastic formations		**LOWER BORROWDALE VOLCANIC GROUP**
		SKIDDAW GROUP

Diagram 3.4 | Formation types.

to recognise on the fells. These highly explosive pyroclastic eruptions allowed the escape of the bulk of the magma in the underground magma chamber. As a result, the domed land collapsed back down along the lines of faults, some of which would anyway have been used as plumbing conduits and, at the surface, fissure vents for the rising magma. The collapsed area of land, often known as a caldera or a 'fault-bounded depression', soon becomes filled with rainwater forming a lake or lakes.

The third and final phase within the caldera-creating cycle saw further, but smaller, pyroclastic and lava eruptions of basaltic, andesitic, dacitic and rhyolitic magma. Many of the fragments of these later eruptions were either injected into the lake from below the water surface or were erupted above water level and then fell, flowed or surged into the lake. These fragments, along with material eroded from previously erupted material above the lake level carried down by streams, form volcanic/sedimentary rocks. As we will see below, many of the sedimentary features looked at in Chapter 1 among the Skiddaw Group are also seen here in the Upper BVG.

The creation of an idealised caldera is illustrated in diagram 2.4, but the gross and rough forces of tectonics seldom provide ideal examples. The Scafell caldera was possibly a half-basin shape, up to 10–15km across. Also, diagram 2.4 oversimplifies the base or floor of the collapsed caldera. This was not a single coherent structure, but was almost certainly badly broken up by faults. Up and down earth movements would have taken place along these faults. This process is known by the unlovely name of 'volcano-tectonic faulting' and would have been accompanied by earthquakes as well as pyroclastic eruptions.

The result would have been a highly complex feature with an irregular base and steep sides. Within the caldera itself, steep drops and steep rises along fault lines separating different blocks would have been created by all this tectonic and volcanic activity. The highly uneven base of the caldera explains why the present day geology is so complex. Patches of rock clearly created from the same magma (according to laboratory analysis) spread like a mad patchwork quilt. Each set of eruptive material fell into whatever depressions were offered on the caldera floor.

Geologists now think that the western edge of the Scafell Caldera is marked by the boundary between the Lower and Upper BVG rock that runs around the western slopes of the Scafell massif, curving to an easterly direction on the northern slopes of Great Gable, Green Gable, Base Brown, Thorneythwaite Fell, High Knott, Watendlath Fell, then on to meet a great north/south fault line at Hause Point on the western shore of Ullswater.

The situation is not as straightforward as this simple description suggests. Some fairly substantial wedges of Upper BVG rock are found a short distance to the west of the main geological boundary. Such a wedge of rhyolitic rocks crops out on the surface north-west of Yewbarrow, largely surrounded by Lower BVG lavas. The narrow band of rock – just 100–200m wide – runs for a distance of 3km from north of Mosedale Beck to near Overbeck Crag. It possibly represents a fault-related boundary of the caldera or perhaps an underground passage leading to a fissure-vent, through which the rhyolitic magma rose upwards but cooled in the passage, now exposed due to erosion.

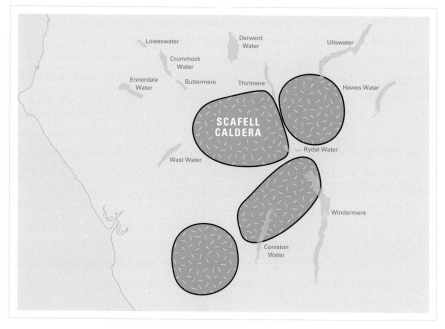

Diagram 3.5 | Borrowdale Volcanic Group calderas. A possible set of calderas and/or 'fault-bounded depressions' created during the Upper Borrowdale Volcanic Group eruptions. The calderas were created one after another in a cycle that usually involved 1) a massive pyroclastic flow (often producing welded tuffs); 2) collapse of land along fault lines creating some form of depression (often with uneven floor) which fills with water into which further pyroclastic fragments fall with other eroded material forming sedimentary tuffs and volcaniclastic sandstone; and 3) a third waning phase often involving some lava and pyroclastic eruptions. The map shows present day lakes to help locate the calderas in relation to present day scenery.

Geologists reckon that another vent was located on present-day Rosthwaite Fell (at grid reference 260 120). However, there were many such vents coming into play during one eruption or set of eruptions then becoming inactive as the centre of action shifted location.

A major fault line in Lakeland runs against the general north east to south-west trend of other major faults. Instead, this one runs north-south right through Lakeland. This is known as the Coniston fault zone. This zone of major faults runs down from a T-junction with the Causey Pike thrust fault to the east of Keswick, via the Thirlmere valley, Dunmail Raise and Grasmere, before it bends slightly to the south-west through Coniston Water. It is thought that the central section of this fault zone marks the eastern limit of the Scafell Caldera.

The creation of the Scafell caldera was followed in time by other episodes of caldera creation, each a short distance away from the last caldera site. Like the main Scafell caldera, these depressions became filled with water and so sedimentary rocks were also laid down (formed out of material eroded from surrounding land including previously erupted tuffs).

Another major caldera was created in the present-day Helvellyn area, with its northern and eastern limits being marked by the Birkhouse Moor fault. This runs from a T-junction with the Coniston fault at Thirlmere just south of Dry Gill, passes just north of the summits of Helvellyn Lower Man, Catstye Cam and Birkhouse Moor, then swings southwards near the present-day village of Glenridding, cutting across the southern tip of Ullswater. From here the caldera limit is not clear.

The eastern boundary of one of these calderas is thought to exist in the eastern fells area where faults around Birkhouse Moor and Hogget Gill appear to have moved up/down relative to each other by several hundred metres. The Birkhouse Moor fault was one site of significant up/down movement of the rocks on either side, by as much as several hundred metres, bringing the earlier Lower BVG rocks directly alongside the Upper BVG rocks in the north and east of the central fells area (see Walk 9).

The Grassguards fault, which runs in a somewhat curved course from just south of the summit of Green Crag to High Tongue (both on the southern flanks of Harter Fell), is thought to be the northern margin of the caldera/fault-bounded depression created by eruptions within the Duddon Basin succession.

All these eruptions and land movements were accompanied by earthquakes. Some of the earthquakes caused deposited material, some of it only semi-solidified, to slide down steep slopes on the floor of the caldera or depression, causing different rock types to be mixed up. Some of these slides were massive, involving great blocks of rock (some as large as 500m long) that slumped dramatically downwards, even embedding themselves in each other. Side Pike and

Blake Rigg (on the eastern flank of Pike o'Blisco) is thought to be such an area (see Walk 15).

The considerable number of eruptions, the different types of rock from each eruption, the land movements along faults, the mass slumping of deposits because of earthquakes and reworking of erupted material from above sea level have all combined to create an incredibly convoluted geological terrain.

It has taken geologists many years to develop an understanding of the mix and work out which rocks were created by which eruptions and in which order. They are constantly revising their views and introducing new categories and names. The Upper BVG cycle is currently divided into three major groups (successions) of rocks, with a dozen or more formations (sub-phases). Within the formations there are scores of 'members' and innumerable variations within each member (this is for the area covered by the 1:50,000 Ambleside geology map; the Keswick and Appleby area maps have slightly different classifications). All in all, it is impossible for anyone but a specialist in sedimentary and igneous geology to keep track of the ever-changing kaleidoscope of names and rock types. I will mention only a handful of these names e.g. Scafell Dacite and the Seathwaite Fells Formation. Trying to identify one formation or member on the ground and differentiate it from another rock type is all but impossible for the amateur.

The hillwalker, however, can usefully focus on what can be seen today on the fells and which has resulted from these eruptions. The most common type of pyroclastic rock is known as tuff. Tuffs are, strictly-speaking, rocks formed from pyroclastic fragments smaller than 2mm in size, but the name is also used more casually by geologists as a generic term for pyroclastic rocks. To simplify matters I will basically refer to three common basic types of pyroclasts: tuffs, coarse tuffs and breccias and also to a special type of tuff known as 'welded tuff'.

The fragments in a tuff are not visible to the naked eye because even although you can see something 2mm in size, this is not really feasible on the weathered/lichen-covered rock surfaces met on the fellside. This means it is often difficult to be sure whether a rock is a tuff or something else (such as a sandstone). Sometimes tuffs can be seen in the form of 'flinty' rock, often a dark blue where broken to reveal a fresh unweathered face, but otherwise with few conspicuous features.

In a coarse tuff, the fragments range from just visible, say the size of a pimple, to about the size of a golf ball (see Photo 3.3). This type of rock is sometimes known as 'lapillistone' or 'lappilli-tuff', but these terms are used inconsistently so I will use the term coarse tuff. If you are within the area covered by the Upper BVG and you see rock which contains small but clearly visible lumps then the chances are it is a coarse tuff.

Photo 3.3 | Coarse tuff, Bowfell.
Field of view 0.6m across.

Photo 3.4 | Breccia, Great Gable.
Field of view 1m across.

Photo 3.5 | Welded tuff, Pike o'Blisco.
Field of view 0.3m across.

Photo 3.6 | Welded tuffs, Crinkle
Crags. Field of view, 1m high.

Breccias are any rocks with fragments bigger than a golf ball. Technically, the fragments should be greater than 64mm. Since few people take a ruler out on the fells with them, I will use the description bigger than a golf ball but smaller than a cricket ball as the dividing line. It will be recalled from the last chapter that lavas can also sometimes produce breccias, so not all breccias are pyroclastic rocks. Pyroclastic breccias come in many varieties (see Photos 3.4, w7.2, w7.5, w7.6, w10.4, w13.12, w13.13 and w15.1). Breccias are probably the most interesting type of pyroclastic rocks and often convey the impression of fiery heat, motion and mass that must have existed at the time of their creation. The larger a fragment, the closer you are to the vent from which it was erupted as bigger fragments travel less far than smaller ones.

Welded tuff is a special form of tuff and is the product of an extremely hot, violent and gassy eruption ('ignimbrite' in the geological jargon) that blasts out a sideways flow of burning clouds of matter at astonishingly high speeds. This type of eruption, as we saw earlier, is commonly associated with the creation of a caldera and so it is not surprising that such rocks are common in the area. The eruption is so sudden and so violent that it rips away lumps of the hardening 'pumice' left in the vent by previous injections of magma (see Photos 3.5, 3.6, w7.9 and w8.4).

In an ignimbrite eruption, these lumps of pumice are subjected to so much heat and pressure that they are flattened and welded directly into the fabric of the other material left behind as the clouds of gas and ash finally settle down. Welded tuffs are often seen on and around the central fells (see Walk 13) and it is a relatively easy type of rock to spot if you keep your eyes open. The key feature is the presence of the flattened pumice lumps which form narrow streaks in the rocks (see Walk 8).

As mentioned above, the creation of the caldera usually leads to the formation of a depression. This leads to rainwater collecting to form a lake. Material from any future eruption is then thrown out into the atmosphere, only to settle in water when it falls back to Earth. Also, material from previous eruptions which fell to Earth outside the caldera is eroded by streams and rivers which then carry it down and dump it in the caldera lake. Sedimentary material therefore collects in the lake and hardens over time into sedimentary rocks.

Traditionally, sedimentary and volcanic rocks are treated as two quite distinct classes of rock ('metamorphic' rock being the third category). Strictly speaking however, the term sedimentary applies to any rocks produced from material that is laid down as in water, in the seas or in lakes, whether from material derived from erosion or volcanic eruption. Where volcanic ash erupts into the air and then falls down into lake or seawater, it forms a 'sedimentary tuff' though this name is seldom used by geologists. Also, any volcanic ash and larger fragments that settle on land are soon subjected to erosion by rain and subsequent run-off in streams and rivers. The eroded material that is carried down to lakes or seas is then deposited as reworked or 'volcaniclastic sandstone'. (Note that the latter label is also applied to sedimentary tuff and the sandstone label is rather loosely used as, in this form, it includes sedimentary rocks with smaller particles, such as siltstone and mudstone).

Photo 3.7 | Bedding in sedimentary rocks within the Upper BVG deposited in lakes within the Scafell Caldera, Scafell Pike, now tilted from the horizontal to dipping from upper right to lower left.

In recent years geologists have reassessed many Lakeland rocks of the Upper BVG and reclassified them from sedimentary tuffs to volcaniclastic sandstones. We need not worry too much about the distinction as all these rocks are considered to be part of the Upper BVG. However, we do need to note the large areas of sedimentary rocks within the Upper BVG area. These are easy to spot and often produce fascinating variations. As we saw in Chapter 1, sedimentary rocks often display features such as bedding, slumping, folding and even cleavage. These features are also clearly present in Upper BVG rocks (see Photo 3.7).

A special feature of sedimentary rocks is 'rippled bedding', where the ripples caused by moving water (seen today on beaches between tides, for example) are recorded in the beds (see Photo w6.4). The ripples indicate that the material was dumped in a stream or river or tidal area where there was some sort of current; the bigger the ripples, the faster the current. A related feature is known as 'cross-bedding' and can be seen in Photos 3.8 (as seen in a slab used in a footpath near Rydal Water) and 3.9 (from an in situ outcrop in weathered form). Here the channels of water shift over time and a new channel cuts across the bedding marks left by a previous channel.

Photo 3.8 (top left) | 'Cross-bedding'. This feature is caused by the shifting of location of channels of water carrying sedimentary material where one bed cuts aross the top of another bed. Rydal Water.

Photo 3.9 (top right | Example of cross-bedding at Lingmoor Fell.

Photo 3.10 | 'Graded-bedding': Pyroclastic fragments erupted into the air which then fell into the water with larger fragments falling to the bottom (Pike o'Blisco, field of view about 0.4m across).

Photo 3.11 | Slumping within sedimentary rocks of the Upper BVG caused by earthquakes dislodging partially dried sediments (Bowfell).

The sedimentary rocks featured in Photos 3.8 and 8.13 are part of what is known as the Seathwaite formation. This is a fairly widespread set of rocks including sedimentary tuffs, coarse tuffs, breccias and volcaniclastic sandstones. However, the illustrated slates (stunningly beautiful when polished) occur only in a few limited beds within the overall

Photo 3.12 | Welded tuffs on Crinkle Crags: part of the Scafell Caldera area, these welded tuffs were erupted from very violent eruptions and then deposited under water in a lake confined within the collapsed caldera. Scafell Pike and Scafell in the distance roughly mark the outer bounds of the caldera.

formation and many other rock types also occur. Previously geologists classified these slates as the Tilberthwaite Tuffs and then as Seathwaite Tuffs before settling (for the time being) on Seathwaite Fells formation. The rocks of the Seathwaite Fell formation were formed from material deposited by eruptions on the land outside the Scafell caldera, removed by subsequent erosion and deposited by streams in the Scafell caldera and in other fault-bounded depressions to the north-east, east and south.

Another feature of sedimentary rocks is 'grading' where heavier and larger material falls to the bottom of a bed while smaller material collects higher up. This feature, known as 'graded-bedding', can be seen in Photo 3.10. Slumping caused by earthquakes dislodging the partly solidified sediments is also often recorded in the rocks (see Photo 3.11).

As mentioned above, the eruptions of the Upper BVG sub-cycle were accompanied by volcano-tectonic faulting and uplift/downshift of land along fault lines. The faults were of all sizes, from very large to quite small. Geologists infer the presence of a major fault from close examination of the rocks and other features. It is harder for the hillwalker to spot a major fault unless there is a clear feature, such as a deep gill marking the line of the fault as in the

Photo 3.13 (top) | Craggy tuff scenery – Coniston Old Man and Brim Fell, centre distance; Dow Crag, right distance; light-coloured, tuff terrain of Swirl How, foreground.

Photo 3.14 (left) | Fault cutting vertically through sedimentary bedding, with the beds on the left slightly lower than those on the right (Side Pike).

Rosset Gill fault. Nonetheless, if you keep your eyes open it is sometimes possible to see faulting and land-movements recorded in the sedimentary bedding of the rocks on a small scale. Photo 3.13 shows such an example from Side Pike. The fault, a fairly small one, runs vertically from top to bottom through these fine beds of tuffs and coarse tuffs. You can clearly see how the beds on the left of the fault are a few millimetres lower than the corresponding beds on the right. The fault must have been activated after the beds were laid down.

I hope this chapter will have highlighted the types of rocks that can be recognised with relative ease in the central fells, such as breccias, coarse tuffs, welded tuffs, forms of bedding, folding, cleavage and even faulting.

A really good way to see a wide variety of examples of these different types of pyroclastic rocks, including breccias, coarse tuffs, tuffs, welded tuffs and other features, is by spending some time mooching around the innumerable rock outcrops between the summit of Pike o'Blisco and the area known as Blake Rigg on its eastern flank (see Walk 15). A considerable mish-mash of pyroclastic rocks has been brought to the surface in one small area by geological chance, especially between Blake Rigg and Long Crag, a feature known as 'megabreccia'. This is the product of massive slumping in this area, of blocks as big as 500m across. An eruption or earthquake must have set off a vast slump of solidified and semi-solidified layers of rock on a sloping caldera floor or following an up/down earth movement. Some of the blocks have actually become embedded in other great blocks.

The average fellwalker will see none of this unless he or she undertakes lengthy geological mapping and analysis of the probable explanations of this mad mix of rocks. The slumping and embedding theory is what geologists have developed in their attempt to explain how the extraordinary hotch-potch of rock could have developed. However, what it does mean is that this is truly a good place for getting an initial overview of a wide variety of pyroclastic rocks, including sedimentary 'volcaniclastic' ones, as well as being a superb walk on a delightfully pretty and compact mountain.

Before moving on to the dramatic mountain-building events that followed the volcanic episode, two more points require mentioning. First, I should add that there are some volcanic outcrops in the area known as the Eycott Group. These appear in an arc on the north-eastern side of the fells, from the Skiddaw massif to near Haweswater (except for a small outcrop in the far south-west of the area). Some geologists argue that these rocks were erupted before the Lower BVG lavas; others include them in the BVG cycle. They are seen on either side of Ullswater, but need not detain us further as they are only a small part of the Lakeland scenery whenever they were originally created.

Second, and more important, another important event occurred at the end of the volcanic cycle of which there is plenty of evidence in today's fell scenery. Once all the magma had been ejected (or cooled as intrusive rock before reaching the surface), there remained in the deep magma chamber a residue of fluids containing the most volatile elements such as chlorides, fluorides and metallic elements. These formed what geologists call 'hydrothermal convective cells'. The hot mineral-bearing liquids were forced up into any cracks such as faults and cooling joints (created as hot rocks become colder). On the way they leached more minerals from the surrounding rock. These liquids cooled to form quartz and similar rocks, many containing metallic elements such as copper and lead in narrow 'veins' (seldom more than a metre wide at the most).

There were several different episodes of such 'mineralisation', some during the volcanic period that we've been looking at in this chapter and several in later times from around 400 million years ago to just 150 million years ago. Mineral deposits are located in both the Skiddaw Group and the BVG areas. Copper, lead and zinc are the most frequently found mineral veins, but antimony, arsenic, bismuth, cobalt, manganese, nickel, tungsten, silver, gold,

Photo 3.15 | Subdued, rolling scenery of sedimentary rocks in southern Lakeland (view from Loughrigg Fell, looking towards Windermere).

Photo 3.16 | Tuff Terrain, the view from above Esk Hause: Great End, far left; Great Gable and Green Ganble, centre left; Base Brown, centre.

baryte, plumbago and iron ore (mainly as hematite but also some magnetite) are also found in some places.

It is probable that at least some of the metallic contents were leached out of the surrounding volcanic rocks by the hot fluids which rose up through faults and cracks in the rocks. Geologists have noticed, for example, that while there are mineral veins found in several different volcanic rock types in the Coniston and Tilberthwaite areas, there are none among the outcrops of the sedimentary volcaniclastic sandstones. Copper-bearing veins are almost entirely located within the part of the Upper BVG area which was affected by tectonic faults and the creation of the major Scafell Caldera. Around Coniston, many veins are found in areas where there are welded tuffs. It's also possible that there were repeated episodes of mineralisation affecting the same veins, which may account for why both lead and copper can be found in the same area or network of veins. We will look at mineral mining and exploitation and the effect on the scenery in Chapter 8.

Well before the episodes of mineralisation occurred, the volcanic era in Lakeland had run its course. A few million years later the volcanic cycle that created Snowdonia in North

Wales began to get under way. Another twenty million years after that and the whole of the Iapetus Ocean had disappeared, its plate fully subducted under the continental plates of Avalonia and Laurentia. During those forty million years, great thicknesses of sedimentary rock were laid down which nowadays outcrop in the south of the Lakeland area, producing a rolling landscape.

The boundary between the volcanic rocks of the BVG and the later sedimentary rocks is marked by the famous Coniston Limestone, a thin band of limestone that runs east to west across the district passing just south of the Old Man of Coniston. Sandstones and siltstones were then laid down farther south. These have produced the subdued scenery of southern Lakeland between Coniston Water and Windermere and to the east and south of Windermere.

It's now time to move on to the next period in the creation of the landscapes of Lakeland. There is now no trace of the ocean plate that once existed between Laurentia and Avalonia. However, the convection forces within the Earth's mantle continue to drive the continents towards one another, setting the scene for a new round of dramatic tectonic effects.

CHAPTER 4

Making Mountains

The convection currents within the mantle continued to drive the two tectonic plates carrying Avalonia and Laurentia towards one another, even after the Iapetus oceanic plate had been fully subducted. The result was a collision of continents and the building of the Caledonides, a massive mountain range which included the mountains of the Lake District.

This mountain range is named after the Scottish Highlands which were created at the same time and in the same way, as were the Welsh mountains, those of Scandinavia and, on the eastern flank of northern America, the Appalachians.

We saw in the last two chapters how the collision of the Iapetus oceanic plate with the Avalonia continental plate had led to subduction of the oceanic plate and the generation of

Photo 4.1 | The view north from Scafell, depicting the rugged heart of Lakeland's mountainous scenery (Great Gable, centre right).

volcanic activity. Something similar had happened in the west, where the Iapetus Ocean was also subducted under the Laurentian continental plate, again leading to volcanic activity.

When the oceanic plate had been fully squeezed underground by the converging continental plates, the continents were crumpled up against one another. The solid layers of rock making up the continents were subjected to enormous forces over a very long period of time. As previously pointed out, the plates move only a few centimetres a year. Over several million years, however, this amounts to considerable distances of 20 to 100 kilometres every million years.

We know from experience that rock is likely to shatter if subject to a sudden force, say from being struck hard with a hammer – it is brittle. However, when rock is subject to long but relentless pressures it can behave very differently. It behaves in a 'ductile' fashion, that is it can fold and even flow (as also happens in the mantle to create convection currents).

The collision of the continents was initially fairly restrained. The outer parts of the continents, the thin continental slope and volcanic 'island arcs' met first and began to fold and crumple. Then the main, much thicker, parts of the continents were driven closer to each other. The rocks piled unrelentingly into one another, crushing and buckling the weaker rocks more than the tougher rocks, but all were affected. The area where the continents meet gets

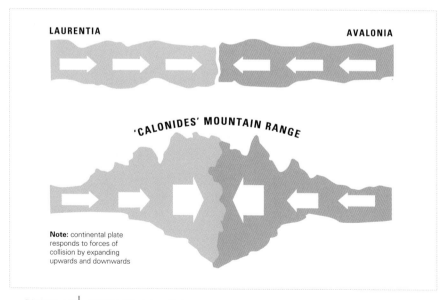

LAURENTIA

AVALONIA

'CALONIDES' MOUNTAIN RANGE

Note: continental plate responds to forces of collision by expanding upwards and downwards

Diagram 4.1 | Continental plate collision.

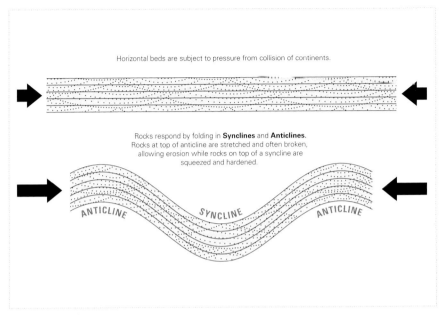

Horizontal beds are subject to pressure from collision of continents.

Rocks respond by folding in **Synclines** and **Anticlines**.
Rocks at top of anticline are stretched and often broken,
allowing erosion while rocks on top of a syncline are
squeezed and hardened.

ANTICLINE   SYNCLINE   ANTICLINE

Diagram 4.2 | Synclines and anticlines.

progressively thicker, pushing up some of the rocks into massive mountain ranges (and also thickening downwards to an even greater extent than upwards). Faults are reactivated or created anew and sometimes, as happened with the Skiddaw Group rocks, deeper layers of rock were thrust up on top of newer layers. The intense pressure and heat caused some melting of the lowest part of the plate to become molten magma that then rose, if at all, only some way up through faults before getting stuck and cooling to become granite (a thick-grained 'rhyolitic' rock). The heat also 'cooked' some of the rocks causing chemical changes, and the enormous pressures generated and imposed on the rock turned fine grained rocks into slate. These last two effects are known as types of 'metamorphism' where rocks are transformed in their state in some way.

The beds of rocks were folded into dips ('synclines') and rises ('anticlines') that can sometimes be detected into today's scenery. Some were shattered (most likely along existing fault lines) and others thrust above one another, even turned upside down. We can see this process of mountain-building happening today before our very eyes in the Himalayas (a result of India impinging on Asia) and in the closer to home Alps (a result of the clash of Africa with Europe).

Powerful pressures were exerted close to the continental edges including, of course, the area that became Lakeland. The sedimentary Skiddaw Group rocks were softer and less able to resist the pressures exerted by the collision than the harder volcanic rocks of the Borrowdale Volcanic Group (BVG). The result was that the Skiddaw Group rocks were intensely folded and broken up with major sections thrust upwards above younger rocks along fault lines. The evidence for this can be discerned in places, especially in Gasgale Gill and on Hopegill Head and Whiteside, the ridge which forms the northern side of the gill. The thrust line is clearly visible at the base of Gasgale Crags, high up the valley side (see Walk 1).

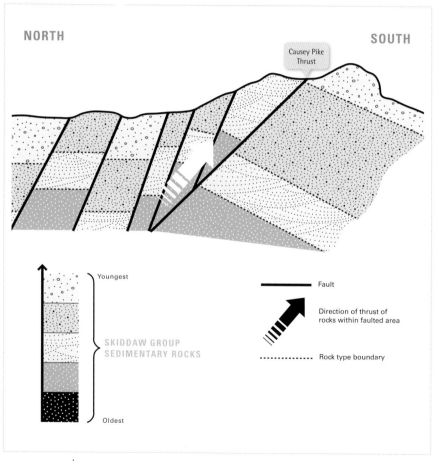

Diagram 4.3 | Folding and thrusting in Skiddaw Group rocks.

Photo 4.2 | Syncline in Skiddaw Group sedimentary rocks exposed on the summit of Hopegill Head.

Photo 4.3 | Small-scale folding in Skiddaw Group sedimentary rocks, Whiteside.

It is also possible to spot in places some of the folds in the rocks and the several anticlines and synclines that run across the Hopegill Head/Whiteside ridge. Indeed, just a few metres west of the summit of Hopegill Head, it is possible to look back up to the summit along the ridge line to see that the top-most rocks are the base of a syncline (see Photo 4.2). The Gasgale Thrust can also be detected further east, cutting through Hopegill Head, Grisedale Pike and the ridge that heads into Hospital Plantation.

Another major thrust line is known as the Causey Pike thrust and runs right across the district. It trends from west-south-west to east-north-east from Crummock Water (on the southern slopes of Rannerdale Knots, Wandope, Sail and Scar Crag) before heading north through the summit of Causey Pike, then resumes a west-south-west to east-north-east trend, eventually petering out near Swinside (south-west of Keswick). Two further thrust

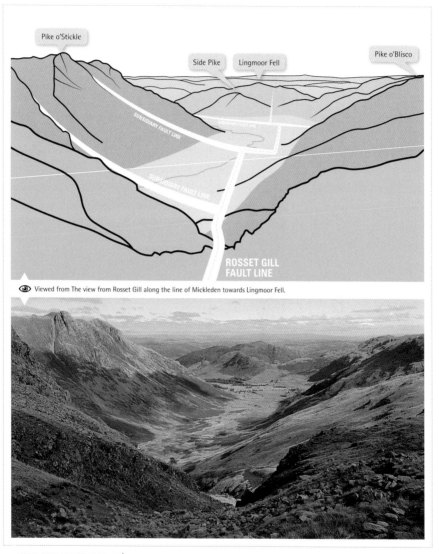

Viewed from The view from Rosset Gill along the line of Mickleden towards Lingmoor Fell.

Diagram 4.4 & Photo 4.4 | Looking along the line of the Rosset Gill fault from Rosset Gill, along Langdale towards Lingmoor Fell. The line of the fault can just be seen as a dark streak on Lingmoor Fell.

faults are found a bit further north-west between the Gasgale Thrust and the present-day town of Cockermouth.

It's worth pointing out that when rocks are folded into anticlines and synclines, the rocks on the top of an anticline are stretched, being on the outer side of a curve, while the topmost rocks of a syncline are compressed (and vice versa for the underside rocks in each case). The stretching of the topmost rocks on an anticline weakens them and cracks them open, giving a head start to the forces of erosion, opening up the lower layers of rock to eventual exposure on the surface. On the other hand, the topmost rocks of a syncline are compressed and thus hardened, making them more resistant to erosion.

This leads to the counter-intuitive effect of anticlines often (but not always) being found in well-eroded areas and synclines often (but not always) being found at the higher points (such as on the summit of Hopegill Head, as seen in Photo 4.2). Both Scafell and Snowdon, the respective highest points in their own regions, are centred close to synclines created by mountain-building forces along calderas created by volcanic activity which took place along those ancient faults - lines of tectonic weakness..

The Scafell syncline trends from north-east to south-west and accounts for the general tilt of bedding that is often seen in rocks in the area (for example on Bowfell as seen on Walk 6). This is a wide and fairly shallow syncline, not a sharp and closely spaced one as seen in the sedimentary Skiddaw Group rocks. This is because the volcanic rocks were more resistant and could better withstand the forces generated by the collision of the continents. Another major syncline, the Ulpha syncline, cuts across often rolling moorlands of Ulpha Fells to the south.

The great pressures exerted by continental collision also caused up and down movements along re-activated ancient faults and on new ones. One particularly important fault, the Rosset Gill fault, runs right against the general trend from north-west to south-east. Its effects on the landscape are obvious in many areas, from Ennerdale, Aaron Slack, Esk Hause, Rosset Gill, Mickleden and across the south-western flank of Lingmoor Fell.

Other major faults run through Eskdale, Upper Langdale, Whillan Beck, Greendale, Greenburn, Grassguards, Baskill, Park Gill, Stockdale, Coniston and Brathay Sideways carth movements of up to 75m arc not unusual (for example, on the Broad Stand fault on the south-east of Scafell) and up to 900m in a few instances (on the Great Langdale Fault). Elsewhere, there have been vertical earth movements along faults of as much as 500m in the granite areas (see below) and among the southern sedimentary rocks.

The faults can often be traced in the landscape over a range of features that together form a longish line. The Eskdale fault runs up through Eskdale, the Wrynose pass and Little Langdale area until it terminates on the Coniston fault zone. It cuts through the granites exposed in the area as well as the Borrowdale Volcanic Group rocks. Some of the Lower

BVG rocks are displaced vertically against one another in the Hard Knott/Wrynose area, with the rocks on the northern side of the fault about 70m lower than their equivalent on the southern side. These faults have often been the site of later deposition of metallic minerals, later mined for copper and lead (see Chapter 8).

The Coniston Fault zone is another of the major fault structures in Lakeland, stretching some 40km from north to south. Major features are spread along the fault zone, including the valleys of St John's in the Vale, Thirlmere (rising to a high point at Dunmail Raise), the valley of the Rive Rothay to Grasmere (another high point at the shoulder between Loughrigg Fell and Silver Howe) and down to and beyond Coniston Water. This major fault zone is believed to have been 're-activated' around 250 million years ago, becoming the site of later earth movements.

The slightly curved Grassguards fault runs from just south of Grassguards Gill, Long-house Gill and Brown Pike. This is thought to be the northern limit of the basin created by the final eruptions of the Upper BVG rocks, the Duddon Basin Succession. The vertical movement along the fault is over 500m, with the rocks to the north of the fault line being lower than those to its south.

As well as the major faults, there are scores of minor ones where earth movements have been less catastrophic, but have still contributed to the shaping of the present day landscape, resulting in minor crags and shallow hollows. These often occur in complex patterns as earth movements along faults have shunted softer and harder rocks into complex arrangements. The faults themselves have also been exploited by forces of erosion, especially ice, to create gullies and valleys.

Photo 4.5
Piers Gill on
Scafell Pike.

Photo 4.6 | Tilted rocks seen on Bowfell.

One consequence of all these earth movements along fault lines has been to smash the rocks on either side of the fault and to create what geologists call a 'fault-breccia'. These can be up to 100m wide and during the ice ages provided weak points along which glacial ice could carve deep gullies into the rock. This is a common feature of the Lakeland landscape. Such gills are seen on Walk 3 (following Piers Gill on Scafell Pike) and Walk 6 (following Hell Gill on Bowfell, Walk 5, descending Aaron Slack between Great Gable and Green Gable on the line of the Rosset Gill fault).

The folding, tilting and faulting during the mountain-building episode have had a considerable effect on the shape of the landscape that we see today. The great tilt of the rocks forming the Scafell syncline are often quite easy to see, such as on Bowfell or the Langdale Pikes (see Photos 4.7 and w7.7).

Another consequence of the forces generated by continental collision, the creation of slate, has less immediate effect on the topography. The very great pressures generated by the collision of the continents squeezed the tiny minerals making up the fragments of deposited particles in sedimentary rocks. This caused the minerals to change their positions so that they all ended up with their flat faces lined up in the same plane; what is called the 'cleavage plane'. These rocks then split easily along the cleavage planes, producing slate. Some of this slate is worth extracting for roofing or building material, but only a small proportion of it.

Photo 4.7 | View of Crinkle Crags and Bowfell. The general tilt of the beds from upper left to lower right is clearly visible, as are deeply incised glacial meltwater channels along fault lines.

The finer the mineral grains (as found in mudstones), the more effectively that good cleavage planes are created and the more likely the rocks are to produce good slate. Slightly bigger mineral grains (as found in siltstones) are less likely to produce good slate, but may well still display cleavage. Coarser mineral grains (as found in sandstones) may not produce any cleavage at all.

I wrote a few paragraphs above that the creation of slate has a less immediate effect on the landscape than thrusting, folding and faulting. However, slate does have consequences for scenery in the longer term. First, cleavage planes in general tend to provide a form of weakness which can be exploited by the forces of erosion. Second, slate is attractive for human use and has led to the development of slate quarries which have left pretty conspicuous effects on the landscape and in the style and colour of the buildings of the area. As the extraction of saleable slate usually involves the quarrying of up to nine or ten times as much unusable slate material, it also involves the creation of large waste tips near slate quarries and mines.

It is worth noting that slate is indeed mined as well as quarried. It is often assumed that a mine in the Lakeland area must be the site of metal extraction, but this is not necessarily the case. Slate is, or has been, mined at Honister and on the southern flanks of Loughrigg Fell, Coniston and elsewhere. Several old slate mines are passed when walking along Loughrigg Terrace, leaving only waste tips and bigger or smaller mine entrances. The waste tips are worth looking at (with very great care if you are tempted to clamber up onto them) for small lumps of the beautifully bedded rocks of the Seathwaite Fells formation.

The story of the collision of the continents does not end there. Underlying the whole area are believed to be substantial sheets of an intrusive rock: granite. The biggest concentrations of granite are thought to have been intruded about 450 million years ago. Some smaller areas (such as the granite around Shap) were probably intruded later, towards the end of the mountain-building events, possibly when the lowest extension of the continental plate during downward thickening of the plate was turned into molten magma.

Granite is a large-grained rock, a feature that arises because of its slow cooling low down in the Earth's surface. It is also a rhyolitic, light-coloured rock (often pinkish in colour). Although such granite is believed to lie underneath the whole of the Lake District it is only exposed in a few places, mainly to the west (see Walks 11 and 14). Where it does crop out, it generally produces a low, rolling moor-like landscape that is found in places such as Ulpha Fell and Birker Fell.

The presence of granite under much of the Lakeland area is inferred from 'gravity anomaly' readings performed by geophysicists with sensitive gravity measuring devices. The results led

Photo 4.8
Looking out
from a slate
mine entrance
on Loughrigg
Fell, now closed
to the public
because of
recent rock falls.

them to believe that the granite formed a single massive blob (a 'batholith') underlying the whole Lakeland area. This presented them with a problem, in that this blob would have had to have been far, far greater in mass than all the erupted material. However, according to the theory of volcanic systems and continental plate collision, this could not really be the case as granite is thought to be produced in comparatively small volumes at the end of the cycle, not in vastly greater amounts.

The interpretation of the magnetic anomaly readings has been revised and it is thought today that the granite consists of separate sheets intruded between layers of rock ('laccoliths'). The combination of the sheets plus their sandwich filling of sedimentary or volcanic rocks are now thought to give the same type of gravity anomaly reading as a single granite mass.

The mountain-building era lasted for several million years, eventually drawing to a close some 300 million years ago. By then, the area had been shaped into part of the Caledonides. This was a massive range of mountains, at least as high as the Alps and possibly even as high as the Himalayas. The very process of crushing and twisting the rocks into such mountains had, at the same time, the effect of opening the rocks up to erosion.

Photo 4.9 | Granite outcrop, Low Rigg; High Rigg in the right background.

Photo 4.10 | Birker Fell: granite topography with tuff terrain of the central fells in the distance.

In the millenia since then, the area has at times been above sea level and sometimes below it. When it was above sea level, it was subject to erosion. When it was below sea level it was subject to deposition of sediments and thus, over time, to the creation of new rocks.

Few traces of those new rocks are left (except for some mineral deposits in fault zones) as they have been fully eroded away. For instance, it is believed by many geologists that during the era known as the Cretaceous, there must have been such widespread deposits of sediments that the whole of Britain would have been covered by chalk. This white rock only now outcrops in southern and eastern England and, if geologists are correct, it has been entirely eroded away from the rest of Britain including the Lake District. The same may also apply to other rocks.

The overall course of events has been the erosion of rock in the area, slowly exposing the deeper and deeper layers of rock, uncovering the ancient layers of the Borrowdale Volcanic Group rocks and the even earlier sedimentary rocks of the Skiddaw Group. The net effect of earth movements and erosion has been to expose, along the surface of the Earth, a slice through what was originally a vertical layer of rock, bringing to our attention the oldest through to the youngest rocks of the area.

This process took place about three hundred million years. It is of course impossible to appreciate fully what such long periods of time actually mean. Castlerigg stone circle is about 5,000 years old. At roughly the same time, stone axes were being hewn out of the volcanic rocks of the Langdale Pikes. Both are 'pre-history', reaching way back before even any retained folk memories let alone written records. However, 5,000 years is nothing, a mere momentary twinkle, compared with 3 million years (let alone 300 million). The net result is that we really have little or no idea what the landscape looked like a long time ago.

The last ice age ended about 10,000 years ago and was only the latest of a long run of ice ages that stretches back somewhat more than a million years. We will look in the next chapter at what the ice age did to the landscape and at the multiplicity of features that it produced. However, as to the original shape of what the glaciers did their work on, we have really very little to go on.

A few hints are gained from the gentle curving slopes of the eastern flank of the Helvellyn range or the summit area of High Street. These are probably surviving pre-glacial surfaces and suggest that the millions of years of erosion since the events described in the last chapter and this chapter left a rounded upland plateau area with gentle slopes and shallow valleys. These dips probably followed the course of present day valleys by and large. However, the scenery was probably nothing like the landscape we know and love today with its sharp plunging slopes and deep glacial trenches.

CHAPTER 5

Shaping Summits

As we saw at the end of the last chapter, the pre-glacial landscape (the product of some hundreds of millions of years of rock creation and rock erosion) was possibly a gently curved highland area with drainage via V-shaped river valleys radiating out from the centre of the higher area.

The landscape we have inherited at the end of the ice ages, the result of several hundreds of thousands of years of glaciation, is a much more dramatic one. Features include sharp ridges, deep U-shaped trenches and the atmospheric 'corries' that sit halfway up mountain sides, sometimes with a tarn impounded in the bowl of the corrie. It is to the ice ages that we owe the variety of the landscape that we love and cherish. Narrow ridges such as Striding Edge and Sharp Edge, as well as glacial corries holding lakes such as Blea Water and Stickle Tarn, for example, rank among Lakeland's most popular spots for visitors. All are the product of ice gouging away at the rock.

After hundreds of millions of years of sedimentation and a few million years of volcanicity, followed by a few more tens of millions of years of crumpling and heating while being crushed into high mountains, the tough resistant rocks then managed to survive another three hundred million years. There were quite probably several ice ages during those long

Photo 5.1
Sunset reflected in a glacial tarn suspended on the blunt, glacially-scoured ridge between Grey Knotts and Brandreth.

periods of lost time. However, the most recent ice age, which began about 2.5 million years ago, has left plenty of evidence for all to see.

It was not a single long and consistently cold period. Rather, there were cold icy periods and intervening warmer times when the ice retreated to the polar areas, before spreading its frozen fingers out once again to encompass land much farther south.

Geologically speaking, we are now thought to be in the middle of an inter-glacial period (notwithstanding any effects that may be caused by human-induced climate change).

The most intensely cold periods occurred in the later part of the ice age, between 75,000 and 15,000 years ago. During these cold periods ice covered most of Britain, including the Lake District, before it retreated once more to the polar and high mountain areas as temperatures started to rise. There was a short-lived blip about 10,000 years ago when there was a resurgence of glaciation during which corrie glaciers were active in mountain areas such as the Scottish highlands, Wales and Lakeland.

In a comparatively short time (seen in the light of the millions of years taken to deposit sediment and to build mountain ranges) the ice made an enormous impression on the fells. However, it is important not to lose sight of the fact that it is the tough, resistant rock that is the key to our much-loved landscape. Elsewhere, in the land that borders the Lake District to the north, east and south-east, less resistant rock has been eroded into much more subdued landscapes of rolling hills, wide gentle valleys and flattened uplands. The tough rocks of Lakeland resisted the destructive forces more successfully and were carved into the stunning shapes we see today. The greater resistance, and thus greater height of the remaining mountains, meant that the summits existed in a colder environment and were thus more likely to see glaciation. Resistance therefore generates more glacial action in the form of deeper valleys and corries.

As mentioned above, there was quite likely to have been a more or less 'radial' drainage pattern in the pre-existing landscape, with the rivers draining the area in all directions from the high central areas to the edges, rather like the spokes of a crumpled bicycle wheel. These rivers most probably began the process of cutting the same valleys that exist in deeper form today.

The pre-glacial radial drainage pattern is believed to have resulted from 'doming' or uplift of the land into a rough dome shape during the last 65 million years. Younger rocks which lay on top of the older rocks were then eroded, exposing the volcanic and sedimentary rocks we have looked at in previous chapters. Repeated periods of glaciation in the last 2.5 million years then set to work on the landscape.

Photo 5.2

The summit plateau of Helvellyn in the frost, looking towards Nethermost Pike and Dollywaggon Pike (a possible remnant of the pre-glacial surface).

It is difficult for geologists and 'geomorphologists' to say for certain what the particular effect any of the glacial periods had on the area. It is seems safest to assume that the glacial features we will look at were the product of a series of glacial periods, rather than only the most recent. In each cold period, ice would have re-occupied valleys or other features initiated in earlier periods and deepened them further.

There are many different opinions among geologists about how the ice affected the Lakeland landscape and what follows is a tentative selection of some of those opinions. Geologists believe that during the maximum stage of the last ice age, from 25,000 years ago to about 15,000 years ago (the 'Late Devensian'), the entire area was covered by an ice cap or ice sheet which was as much as 700m thick at its centre. The difference between an ice cap and an ice sheet is only that of size; different geologists use different terms when discussing the Lake District, so for consistency here I will stick with the smaller of the two types: ice cap. A much bigger ice sheet, or set of interconnected ice sheets, also covered nearly the whole of Britain at the height of this period with a major ice sheet flowing out from Ireland to the west. The Lakeland ice cap seems to have consisted largely of ice created over the highest ground of the area itself, with the larger ice sheets moving south around the Lakeland ice cap to the east and the west.

An ice cap develops when the amount of snow falling in winter exceeds the amount of snow that melts during the following summer. If this situation persists, then more and more snow accumulates each year. Eventually the snow is transformed into ice by the pressure of its own ever-growing weight. It first becomes compressed into what is known as 'firn ice', then becomes much more dense, turning into 'glacier ice' or 'blue ice'.

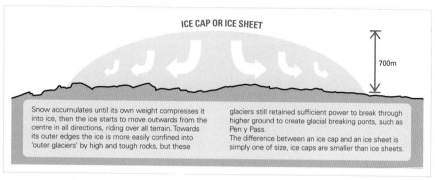

ICE CAP OR ICE SHEET

700m

Snow accumulates until its own weight compresses it into ice, then the ice starts to move outwards from the centre in all directions, riding over all terrain. Towards its outer edges the ice is more easily confined into 'outer glaciers' by high and tough rocks, but these

glaciers still retained sufficient power to break through higher ground to create glacial breaking ponts, such as Pen y Pass.

The difference between an ice cap and an ice sheet is simply one of size, ice caps are smaller than ice sheets.

Diagram 5.1 | Ice caps and ice sheets.

Eventually, the weight of the overlying ice, firn and snow causes the lowest layers of glacier ice to deform and to move outwards away from the pressure. Solid ice thus flows away from the centre in any direction it can, depending on surrounding obstacles, digging down into any weaknesses as it goes. The ice flowing out from the centre of the Lakeland ice cap would have used any suitable channels, forming, where possible, thick 'outlet' glaciers or more confined 'valley' glaciers as necessary.

Geologists can determine the general direction of ice movement from scratches or 'striae' on exposed rock surfaces, scoured into the rock by lumps of rock from higher up, then held in the frozen grip of the moving glacier. These can be seen on many ice-smoothed outcrops.

The centre of the ice cap was the high land of today's central fells – the Scafell/Bowfell massif and the Seathwaite Fell/Glaramara/High Raise area. Some of the higher summits may, for some time at least, have stuck out above the ice before and after it reached its maximum extent. However, when the temperatures were at their coldest the entire area was covered so even the highest summits experienced direct glacial erosion of some sort.

As a result there are few sharp, pointed mountain summits in the area. Many of the fells have a sort of summit plateau around them with sharp slopes only where glacial valleys or corries have been carved. These rocky summit areas are the product of 'periglacial' action from when the tops stuck out above the ice and also after the ice had melted but it remained cold. Repeated 'freeze-thaw' cycles caused water caught in cracks and joints in the exposed rocks to freeze, expand and then thaw. This caused the rocks to split apart into the boulder fields and 'tors' that we see today.

The summits of Scafell Pike, Scafell, Bowfell and Pike o'Blisco are typical examples. Each has a summit plateau area bounded by glacial cliffs and topped by great boulder fields, ex-

Photo 5.3 (top) | 'Outlet' glacier from Greenland ice sheet. Photo John Simmons, Geological Society of London.

Photo 5.4 (middle) | Glacier formed within the crater of dormant volcano Kilimanjaro, illustrating the relationship between height and glaciation. Photo Jim Young.

Photo 5.5 (bottom) | Glacial scratches or 'striae' on exposed rock surface, Bowfell.

posed outcrops and tors. The summit plateaux of Scafell Pike and Scafell are particularly rocky, Scafell Pike being a mass of scattered boulders. These peaks are surrounded (and separated) by some of the most dramatic crags in Lakeland.

As the redoubtable Wainwright put it in his pages on Scafell Pike: 'Crags are in evidence on all sides, and big areas of the upper slopes lie devastated by a covering of piled-up boulders, a result not of disintegration but of the volcanic upheavals that laid waste to the mountain during its formation.'

It is with enormous trepidation that the newcomer guidebook writer dares contradict the great AW, but here he does seem to have got it the wrong way round. The boulders are far from being the 'ejecta' of those ancient volcanoes, although the rock itself was of course originally created by the volcanoes. The boulders are indeed the product of disintegration of the rock, due to freeze-thaw action on the exposed rock in very cold times. The crags themselves result from great glaciers that carved deep into the mountains.

Bowfell, when seen from a distance, often looks distinctly plateau-shaped, slightly tilted and rising to its actual summit. The tilting is particularly evident as you ascend or descend from Three Tarns to the summit, with beds clearly visible in the rocks on either side of the track, and most especially when you pass the 'Great Slab'. Pike o'Blisco looks cone-shaped when seen from Great Langdale. However, when you approach its summit you climb quite clear rock steps that represent strata of rocks that form the summit plateau area.

The eastern fells are fairly subdued in comparison with the central peaks. High Street typifies the scenery here, with smoothly shaped summits often with flat tops which are littered with periglacial boulder fields. There are a few tors (such as on Thornthwaite Crag, Walk 9) but there are far fewer rocky tops than, say, on mountains in the region of Scafell Pike or Bowfell. The smooth tops are probably survivors of the pre-glacial landscape (see photos 5.2 and w9.3).

We've seen in previous chapters that the sedimentary rocks of the Skiddaw Group, forming the northern and north-western fells are different from each other and from the volcanic summit areas. Smooth ridges and isolated eminences typify the north-western fells and sinuous curves of the Skiddaw massif. However, the summit of Skiddaw is fairly flat and the highest peak in the north-western fells, Grasmoor, again features a flattish summit plateau littered with boulders. These summit plateaux are also thought by geomorphologists to represent pre-glacial surfaces.

In large parts of the central fells area, even the overall topography of the upper reaches is often restrained. This is often not what you would expect from the valley floors, most notably in Great Langdale, where there is a fantastic view of the pointed tops of the Langdale Pikes, especially the delightfully named Pike o'Stickle. The pikes appear as the sharp tops

Photo 5.6

Boulder field and tors created by freeze-thaw action in rhyolitic pyroclastic rocks on the summit plateau of Swirl How.

of mountains and give every impression of having similar looks from the other side. This is an illusion, however, that is shattered on the first occasion that you ascend to the summit plateau below the Pikes.

The land rolls off roughly, rising steadily towards Thunacar Knott and High Raise, offering a wholly unexpected view that is not what seemed likely from below. The Pikes are much less impressive from this side, mere minor eminences among the higher but less immediately scenically attractive northern summit plateau. However, after the initial surprise, this is a landscape that grows on you, especially if you can imagine it covered by masses of ice flowing roughly northwards, gouging out the rough landscape you see before you (Walk 7).

The summit also offers fine and fairly easy extensions of walks from the area of the Pikes northwards to Thunacar Knott and High Raise. The hard work of climbing is largely

done and the ascent that remains is on relatively gentle slopes. The crowds that flock to the Langdale Pikes are generally absent here, few bothering to venture beyond the immediate area. Those who do trek northwards are rewarded with a feeling of immense openness and refreshing views.

The ice flowed out from the centre of the ice cap in pre-existing valleys, but also scoured the fell tops at various heights, riding over many flattish and very boggy summits such as the High Seat ridge. While High Street is very popular, some of these are not heavily walked fells as the going is rough and wet and the views are limited by the broad summits.

Elsewhere, the ice created a much more fascinating landscape of hummocky, broad ridges or low fells such as Blea Rigg/Silver Howe, Grange Fell, High Rigg, Haystacks, Lingmoor Fell and Loughrigg Fell (see photo 5.11).

This is known as 'knock and lochan' topography, after the classic landscape of north-west Scotland. It is the result of mass ice movement plucking at weaknesses in the rock, such as slightly softer layers of rock between harder layers. We have seen (Chapter 2) how the 'trap topography' of lava sheets interspersed with thin pyroclastic flows produce the knobbly landscape of High Rigg and other lava areas.

The hummocky fell tops are another version of the same effect, with ice working at a boundary or weakness in the rock to create a small depression (which may become filled with a lake) surrounded by small prominences. These areas can present the hillwalker with a navigational nightmare if the mist descends while in the middle of such terrain. Determining which little pattern of contour and crag on the map represents the real-world pattern of dip and rise, bog and rock outcrop in front of you can be perplexing. It can be difficult to

Photo 5.9

The Langdale Pikes
from Great Langdale.

Photo 5.10

The view north from Pike o'Stickle towards High Raise, centre left, over the unexpected summit plateau.

distinguish between hummocky terrain that is due to erosion by ice and hummocky moraines that have been created by glacial deposition (see photo 6.8).

For inexperienced walkers this sort of terrain can even present a challenge in good light. Once, while descending along Blea Rigg to Silver Howe hoping to get back to Elterwater before dark, I met a woman heading uphill. She told me that she was going to Chapel Stile. Having kept a close eye on my position on the map I was sure that Chapel Stile was in the opposite direction from which she was heading. Although she had no map, she insisted at length that the track down to Chapel Stile was farther uphill as she had walked up from the direction I pointed to and had seen no track. In the end I simply said that I was going my way and that she was welcome to accompany me if she wished, but I would not go with her back up the hill just as it was about to get dark. Fifteen minutes later we met the track down and I led her to Chapel Stile even though I had originally wanted to continue to the pass beyond Silver Howe. When we arrived in the village she turned and walked off without even a goodbye.

Good map-reading is extremely useful if you want to minimise the chances of navigational mishap while exploring these classic Lakeland landscapes. Although they are often not very high in altitude, they often demand constant up and down walking and can therefore be surprisingly longer and tougher than one might expect from glancing at the map. However, these are fascinating landscapes which offer interesting walking for those keen to appreciate the variety of Lakeland landscape. Areas such as Grange Fell or High Rigg often provide good fellwalking and good views when clouds sit on the higher fells. They are probably best left alone in the mist, however, unless you relish a challenge.

There seems to be no overall pattern in the distribution of humps and dips when you are among them; they seem utterly random. However, with the help of a geology map, an explanation of the ups and downs may become clear. It helps to climb to a slightly higher viewpoint where you can look over the whole area, such as the summit of High Rigg which is near the northern end of the summit plateau and provides a fine view back south over the numerous ups and downs. Once you have spent a while looking at the overall view in front of you, it is possible to see some sort of order in the hummocks, with a clear pattern of the higher points standing in lines that cut diagonally across the plateau from left to right, getting farther away to the right. One particularly big line of higher points limits the view. Each line represents an outcrop and each dip represents some linear rock boundary that has been eroded (see Photos 5.11 and w12.7)

Another illustration of the variety of 'knock and lochan' topography can be found, for example, in the area of Angle Tarn on the lower flanks of High Street (see Photo w9.1). Here a mountain tarn has developed out of rainwater collected in a depression carved by glacial ice in this flattish area (made of andesitic lava) bounded to the north and east by tougher rock (a different type of lava to the east and pyroclastic rock to the north). This Angle Tarn (there are two more Angle Tarns in Lakeland) is not the result of a corrie glacier (see Chapter 7) centred at this mezzanine level on the side of a massive mountain ridge. Instead, it is the product of ice cap glaciers moving across the area, plucking and tugging at rocks, breaking in at any weaknesses and carrying away the rocks that are broken off. There are several well-known tarns in Lakeland which owe their existence to this effect, rather than to the action of a corrie glacier.

Photo 5.11
Hummocky terrain on Redtarn Moss and High Teighton How between Wrynose Pass and Pike o'Blisco, view from Hell Gill Pike.

Photo 5.12 |
(Top) knock and
lochan topography:
Angle Tarn.

Photo 5.13 |
(Right) knock and
lochan topography:
Pike o' Blisco.

Before finishing this chapter, we should glance at two other glacial topics. First, a 'micro-scale' feature. Freeze-thaw cycles can shift and sort pebbles on the surface into polygonal shapes and into groups of different sizes (see Photo 5.14). Sometimes, on slopes, the result is that the stones are moved into lines.

Second we should note a 'mega-scale' feature of glaciation, one that affects the whole area (and beyond). The weight of the ice that piled up in the area during the ice age was so utterly and incomprehensibly great that it managed to press down and deform the 'ductile' rocks in the inner mantle just below the tectonic plate. Since the waning, and subsequent complete melting of all the glaciers in the area, that weight has been removed and the mantle rocks are very, very slowly bouncing back into their original shape, pushing up the mountains above them. This is a phenomenon known as 'isostatic rebound'. However, it is worth noting that the rate of height increase due to isostatic rebound is just about the same as the rate of height loss due to erosion, so the present day heights of the mountain summits above sea level are just about static.

We have seen in this chapter how glacial and peri-glacial action is responsible for shaping the summit scenery of Lakeland. It has produced bouldery summit plateaux and tors with hummocky terrain and tarns nestled into any dip or hollow on slightly lower level surfaces, especially blunt ridges such as Blea Rigg. In the next two chapters we will pay attention to how glaciers carved out the deep valleys and atmospheric glacial corries that characterise much of Lakeland. As we have seen, however, the summits and the areas around them are just as much a product of glacial erosion.

CHAPTER 6

Glacial Gouging

The sheer beauty of Lakeland's mountain scenery is vastly enhanced by the great, narrow lakes and valleys that radiate out from the central fells of Lakeland; Windermere, Buttermere and Ullswater to mention just a select few. The valleys and the lakes are both products of glaciation, where ice has carved the present day scenery out of tough volcanic rocks.

As we saw in the previous chapter, the pre-glacial landscape was probably a fairly smooth dome-shaped area, with 'radial' drainage in shallowish V-shaped valleys. The pre-glacial river valleys had, in many cases, developed along the line of major faults and were considerably deepened over the years of glaciation. The ice flowing out from the ice cap in the centre of Lakeland naturally exploited such weaknesses, driven on by its ineluctable need to move outwards. As the ice cut into the central fell summit plateaus and into the valley sides, fallen rock was transported along the glacier to be deposited elsewhere.

This created the classic 'U-shaped' glacial valleys down which 'outlet' or 'valley' glaciers flowed, carving previously V-shaped river valleys into flat-bottomed trenches. Although geologists use the term U-shape, the glacial valleys of Lakeland are more parabolic in profile with a wide floor and steeply rising sides. True U-shaped valleys can be seen in places such as the Alps and other high mountain ranges where the forces of glaciation are still active.

Photo 6.1 |
Valley glacier, Himalayas.
Photo Ian Smith.

The glacial valleys spread out from the central fells area like spokes on a battered bicycle wheel. However, as we shall see, the eastern fells almost form a distinct drainage area, a smaller scale version of the Lakes as a whole. Despite these imperfections, the overall pattern of the flow of water from the centre outwards is immediately striking when looking at a topographic map of the whole area.

These valleys produce one of Lakeland's most characteristic landscape aspects, exemplified perhaps by the view along Great Langdale (frontispiece photo) or along Wastwater. This latter view was voted Britain's favourite on a BBC TV poll in September 2007 while I was camping at Wasdale Head and researching this book (see photos 6.2 and 6.3). Others valleys, such as those housing Ullswater and Buttermere, are just as fine, so I think it is really impossible to say which is the finest. Each has its attractions. Each is a product of a specific set of circumstances, primarily the rock types, presence of faults and relation to the centre of the ice cap.

Despite their common features (a more or less U-shaped cross-section and sharp, sudden valley heads), there is a great variety to the great glacial valleys. Some are linked by a low col at their heads with another valley, such as at Styhead linking Wasdale to the south and Styhead Gill to the north or at Stake Pass linking Langstrath to the north and Great Langdale to the south-east. When it was at its greatest depth and extent the ice would have crossed such low passes, flowing outwards from the centre of the ice cap in any possible direction.

While many of the valleys are pretty straight, such as Wasdale and Ennerdale, others such as Ullswater and Great Langdale have sharp bends. Some are single main valleys, for example Wasdale, while others such as Borrowdale are combinations of valleys. While many

Photo 6.2
Wasdale and
Wastwater:
the deep, wide
valley was
carved by a
valley glacier.

Photo 6.3
Wastwater,
looking towards
Wasdale Head.

lose their identity and simply end where softer rocks have allowed the valley to widen, such as Windermere, others such as Borrowdale and Ullswater have had to overcome serious obstacles and even to divide. Some have long, ribbon lakes within the narrow valley (Buttermere, Ullswater and Wastwater), while in others such as Windermere, a lake is impounded where the valley spreads out when it reaches softer rocks. In fact, only three of thirteen major lakes (Thirlmere, Wastwater and Haweswater) are situated within the Borrowdale Volcanic Group rocks; the others are located mainly to the north and south where there are softer sedimentary rocks.

In other words the variety and beauty of forms within the glacial valleys of Lakeland is never less than astounding. Each valley is different and each is enchantingly beautiful in its own way.

The most dramatic feature of the valleys is surely the Jaws of Borrowdale (see photo 2.9). Here, the valley glacier draining north from the central fells has carved an immensely impressive way past the barrier presented by tough lava rocks from the Lower Borrowdale Volcanic Group. The source of the ice was farther south on the central Scafell Pike/Great End, Bowfell massif, extending northwards over Glaramara and beyond. The ice moved out from the immensely heavy centre of the ice cap into the Grains Gill (itself fed by two minor gills, Ruddy Gill and Styhead Gill) and the Langstrath valleys.

Styhead Gill is a 'hanging valley', entering the main valley some 200m above the Grains Gill valley floor with a series of waterfalls known as Taylorgill Force, allowing its waters to drain to the lower level. This feature, which is quite common in Lakeland, is caused by the smaller glacier in Styhead Gill deepening its valley less than the main ice flow in Grains Gill. The ice in the main valley may have been deep enough that the ice from Styhead Gill simply merged into the main flow or there may have been a height difference along with an icefall. More ice would have joined from Sour Milk Gill, another hanging valley (noting that there is more than one Sour Milk Gill in Lakeland). From this point, where Seathwaite is now located, the valley takes on the classic wide U-shape.

There are some other interesting glacial features in the Borrowdale valley. To the north-west and south-west of Rosthwaite are two hills that stick out from the valley side, Castle

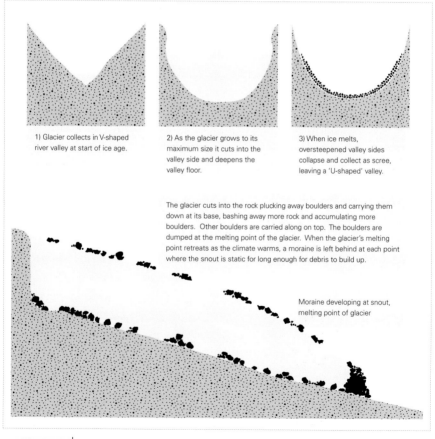

1) Glacier collects in V-shaped river valley at start of ice age.

2) As the glacier grows to its maximum size it cuts into the valley side and deepens the valley floor.

3) When ice melts, oversteepened valley sides collapse and collect as scree, leaving a 'U-shaped' valley.

The glacier cuts into the rock plucking away boulders and carrying them down at its base, bashing away more rock and accumulating more boulders. Other boulders are carried along on top. The boulders are dumped at the melting point of the glacier. When the glacier's melting point retreats as the climate warms, a moraine is left behind at each point where the snout is static for long enough for debris to build up.

Moraine developing at snout, melting point of glacier

Diagram 6.1 | Glacial valley deepening and deposition of moraines.

Crag (High Hows) and High Doat. A channel runs behind each hill at just over 200m. These are thought to have been eroded by streams running parallel to the glacier. These are fairly common features in Lakeland, with some fine examples on Muncaster Fell and Corney Fell.

Farther along, around present day Seatoller and Longthwaite, the ice flow would have been greatly increased with flows joining in from around Honister and Seatoller Fell, from Langstrath and Greenup Gill, and a short distance farther on from Tongue Gill. The combined ice carved out the wide, flat area around present day Borrowdale (and where, after the ice had melted, there was once a lake probably up to 15 or 20m deep). As it travelled farther north, the ice met the constriction of the tough lava rocks at the northern limit of their outcrop. The ice presumably had to rise over the rocks that today stand proud as the Jaws of Borrowdale. Once past that obstacle, the ice carved out a much wider valley, today largely filled with Derwentwater.

The classic glacial valley lake can be impounded by either, or both, of two causes. The first is by 'over-deepening', where the glacier cuts down into the bedrock more deeply where the glacier is constricted between hard ground on either side than it does towards its snout, where its vigour is largely lost and a rock bar is created. This is the case with Wastwater which, at its deepest, is below current sea level. The second is by deposition of rock at the snout of the glacier to form a 'moraine'.

We should note that while some forces are creating glacial valley lakes, others are destroying them. As erosion continues, mountain gills bring down eroded material from higher

ground to be deposited when the stream enters a lake once it meets the valley floor; it no longer has the momentum to transport the material it has carried this far. As this process continues, valley floor lakes slowly get filled in. Indeed, as we have seen, the lake that once existed around present day Borrowdale has been entirely filled in. Also, Derwentwater would once have extended as far back as Grange and the Jaws of Borrowdale. This is easily imagined on a wet day if one crosses the southern end of the lake (Walk 11) on the public footpath. This is shown as land on maps, but the National Trust has had to build up the footpath onto raised walkways to cross the decidedly wet land.

Derwentwater is held in place today around the Keswick area by glacially deposited material forming the curving hummocky ground in such areas. It would originally have formed a single large lake with Bassenthwaite Lake nearly 5km to the north, but again mountain streams from the north-western fells and the Skiddaw massif have dumped enough material to reclaim this stretch as land.

Once beyond the confines of the tough volcanic rocks, the glacier would have widened out considerably. Here a low, rolling landscape is the creation of glacial meltwater channels and glacial deposition of what are known as 'drumlins'. These form the immensely pretty islands seen today poking up through the lake surface in Derwentwater.

Towards the end of the ice age, the glacier would have begun to 'retreat'. This does not mean that the ice flowed backwards. Indeed, the ice still flowed from its source to its snout, even as the location of the snout retreats. It is the melting point that is retreating. This is quite likely to happen in stages, with the snout stopping at one point for a while, then retreating a short distance to another point, stopping and then moving again. At each stopping or resting point, material carried down with the ice would be dumped, forming a 'recessional moraine' (sometimes also known as an 'end moraine' or a 'terminal moraine').

Quite often in Lakeland's glacial valleys you will see complex series of low-rounded, grass-covered mounds arranged along the valley sides, which are clusters of moraines left by the retreating glacier. They cannot be from earlier glaciers as they would have been wiped away, so they give geologists useful hints about the halting way in which the ice age came to an end.

These clusters of moraines, or 'hummocky drift', are not confined to the valley floors. Near Honister Hause there is a high area of hummocky drift. Another example occurs at Langdale Combe (best seen from the eastern end of Rosset Pike). Here the hummocks are distributed in an intense manner filling the entire depression. This allows us to form a picture of the later stages of the glacier, which originally came from the higher land above the combe and then down into Mickleden and Langdale (although Langdale Combe is also an 'immature' glacial corrie).

Photo 6.6 | Moraine complex seen in Mickleden at the head of Great Langdale.

Photo 6.7 | Hummocky moraine terrain in Langdale Combe.

Having looked in some detail at Borrowdale, let's go round the great valleys and lakes in a counter-clockwise fashion and look at some of the distinctive features of each. Buttermere has its joint head at Gatesgarthdale Beck (Honister Pass) and Warnscale Beck (between Hay-

Granite

Pyroclastic rocks of Lower Borrowdale Volcanic Group

Sedimentary rocks of Skiddaw Group

Lavas of Lower Borrowdale Volcanic Group

Viewed from Fleetwith Pike.

Diagram 6.2 + Photo 6.8 | The Buttermere valley from Fleetwith Pike: High Crag and High Stile, left.

stacks and Fleetwith Pike). The two streams of ice united at what is now just above the start of Buttermere Lake. ('Mere' means lake making the latter term redundant in most cases, but as Buttermere is often used as a name for the valley and for the cluster of farms and hotels between the two lakes, it is sometimes justified for aiding clarity of meaning).

As it passed down the valley, the glacier smashed through the rock that linked the pre-glacial heights on either side. It chopped off the spurs of rock that today stand as great crags high above the lake, and which provide such fulfilling opportunities for fellwalking. A run of three or four glacial corries stands high on the south-western side of the lake, best seen below the High Crag, High Stile and Red Pike ridge, in Burtness Comb and in the adjacent bowl enclosing Bleaberry Tarn. Additional corries on Grasmoor contributed more ice further down the valley.

Buttermere is now separated from Crummock Water by a kilometre-long stretch of land. This was built up from material deposited by the streams coming down the fells on either side, and which concentrate on this central point (including another Sour Milk Gill).

Ennerdale is straight and long, running along the line of a major fault. The valley head is centred on the fault line where it passes between the popular peak of Great Gable and its smaller neighbour, Green Gable (this fault, as we have seen, continues through Aaron Slack, Rosset Gill and Langdale to Lingmoor Fell). There is an interesting feature just below the valley head of Ennerdale. The stream Tongue Beck runs from just below the summit of Green Gable to join the main valley as a small hanging valley 1.5km lower down, separated by a spur known as Tongue.

The minor valley holding Tongue Beck runs parallel to the lower valley and has been carved by the ice on a subsidiary fault of the main fault. The subsidiary fault branches off from just below the Black Sail YHA hostel and then runs up to Green Gable along the course of the stream. The stream has been diverted from its minor valley to the main valley, some distance before the point where the fault branches off, by the presence of a boundary between two rock types. This boundary has juxtaposed a comparatively harder rock at this point across the fault line, presenting an obstacle which the stream had to go around. This feature can best be seen from Black Sail Pass between Kirk Fell and Pillar (see Walk 4).

The glacier here has carved a deep channel through the fells, using the weakness of the major fault, leaving immensely impressive crags on both sides of the valley but nowhere more conspicuously than on the northern flanks of Pillar. Once past Pillar, the fells are less restrictive and the valley loses its trench-like shape for a few kilometres. Farther along the valley the fells once again presented obstacles to the movement of ice, producing more fine crags on Bowness Knott and Crag Fell. Beyond this point the fells fall away. Here the glacier would have widened out, becoming what is known as a 'piedmont glacier' (see Photo 6.9).

An idea of what this would have looked like can be gained from looking at the shape of Ennerdale Water on a modern map and, in particular, the wide part of the lake beyond Crag Fell and Bowness Knott. The glacier would also have spread into such a lobe shape just like the lake. In fact, the lake once extended beyond its present position as far as the area between Broadmoor and Howside where there are morainic deposits.

The same is true of Wasdale, the next valley round. Wastwater once extended much farther up the valley to Wasdale Head, though it is impounded at its south-western end by a large moraine area. In the case of Wastwater, there would have been a lake in the valley even without the moriane blocking its south-western end. The glacier has 'over-deepened' the valley below the current lake so that the bottom of the lake is now below current sea level at its deepest point. Solid rock rises beyond the lake to impound its waters. This overdeepening happens most effectively where the glacier had been confined in a narrow space between tough rocks, thus the deepest part of the Wasdale glacier is found between Middle Fell and Illgill Head.

The Wastwater Screes are a famous sight, perhaps because the narrow road on the other side of the lake is packed with motorists during the summer months looking for beautiful views and undemanding walks. However, the screes present a considerable challenge to those not accustomed to fellwalking who want to walk all around the lake. The experienced fell-walker should have no real difficulty crossing the screes (although care is of course needed) which cover much of the south-western side of the lake. Some of the scree is pretty massive and it is this section which presents the only difficulties. If you do walk here (see Walk 14) you

Photo 6.9
Ennerdale.

Photo 6.10 | The Wastwater Screes.

may notice that the screes continue to descend under the level of the lake for as far as you can see; what we see is only the upper parts of these great masses of scree.

The scree would all have fallen from the cliffs after the ice had melted. This happened because the glacier cut through the cliffs low down, but also acted as a sort of buttress holding the undermined higher cliffs in place. Once the ice had gone, the over-steepened cliffs succumbed to gravity and lumps, large and small, peeled away from the rock face and tumbled down to collect on the valley floor. In post-glacial times, freeze-thaw action would have worked on cracks in the remaining crags. Freezing and expanding, then thawing and contracting water in repeated cycles would have prised the rocks apart and added to the collection of scree building up into great piles high above the valley floor. As new blocks or smaller lumps fell off, they either collected on the scree or tumbled further down depending on the angle of the slope; scree has an angle of stability of about 40 degrees. Vegetation starts to colonise scree almost as soon as it becomes stable (though it will take longer where the lumps of rock are quite large as it is more difficult for the gaps between the rocks to be filled.

The glacier in Wasdale retreated all the way to Wasdale Head where it deposited a series of moraines in its last stages around Wasdale Head farm, forming a very distinctive feature seen up close if you walk up this valley or in perspective from the fells on either side. Other

Diagram 6.3 + Photo 6.11 | Deeply cut gills, the product of powerful glacial meltwater streams acting on weaknesses created by faults and rock boundaries

moraines can be seen in the Mosedale valley, where a tributary glacier joined the main glacier. As the main glacier retreated it would have split into two separate glaciers at the point where the Mosedale valley joins the Wasdale valley just above the hotel at Wasdale Head (this is not the only Mosedale in the area, another near Eskdale is mentioned below).

The deeply etched gills on the flanks of Scafell Pike above Wasdale Head, including the highly dramatic Piers Gill (Walk 4), are also glacial features. They have each been carved along a fault, thus giving them their linear character. Rocks on each side of the fault were pressed into each other under great tension, but eventually the earth on either side moved under great tectonic pressures, either lengthways or up/downwards. As the movement took place, the rocks on either side of the fault were crushed as they moved relative to each other forming a 'fault breccia'. This area of weakened rock was then prey to the forces of glacial erosion.

The deep gills were probably created by sub-glacial streams moving under great pressure and carrying boulders which acted as battering rams, prying away at the weakened fault breccia. A very good view of the set of such gills on the flanks of Scafell Pike is had from the flanks of Great Gable, as seen in Photo 6.11 (Walk 5).

Eskdale and Lingcove are more complex than the valleys we have looked at so far, and from most of those mentioned below. They form one of the most atmospheric of Lakeland's valley complexes, but it is a long, hard walk especially if you intend to climb one of the fells at the head of the valley, Scafell, Scafell Pike, Esk Pike or Bowfell, and return the same distance.

However, it is worth walking up Eskdale as an outing in its own right, perhaps with a trip to Border End and Hard Knott, especially if you go beyond the lip into the hanging valley of upper Eskdale where the river is squeezed between crags known as Green Crag and the wonderfully named Throstlehow Crag. Once above the lip, you are confronted with the low crags of Scar Lathing before rounding into the wide, boggy bowl of Great Moss. The surrounding crags, especially on the left running up to Scafell and Scafell Pike, are sensational. The hummocky moraines attract less attention than the great rock faces but, as remnants left over from the last days of the great ice ages, are worth noting.

However, an excellent overview of upper Eskdale and the surrounding area is had from the south summit of Scafell Pike, from Esk Pike and also from Bowfell. Border End and Hard Knott also provide fantastic views and represent easy ways to see the upper Esk valley from the other direction (easy because the ascent from the top of Hard Knott Pass is pretty short). The atmosphere is not the same from these higher vantage points, but it is almost as if you are looking over a scale model of the valley complex. The most noticeable feature of the upper valley from the Scafell massif is surely the sudden 90 degree bend made by the river just before it reaches the lip of the main valley. From here, and by looking at the map, it is clear that the section of Eskdale into which the river falls is really in more of a direct line with Lingcove Beck, which also falls into the same main valley on the eastern side of Throstlehow Crag. However, from the map it is also evident that Mosedale is a continuation of the Lingcove Beck valley. Only a small rise in the land, a sort of rock bar just north-west of Black Crag, diverts the Lingcove Beck into Eskdale instead of Mosedale.

Presumably the ice from Lingcove did originally run into Mosedale, but some of it diverted into Eskdale as it was widened and the divide between the two valleys was reduced. Equally, the ice from upper Eskdale cut across the craggy land south-west of Green Crag at an early stage. It also became deep and wide enough to meet the ice that flowed in from Lingcove, forming a joint glacier with greater power that then cut deep down to create the main valley; this is where Eskdale takes on the classic U-shape profile. The streams have followed this course since the glaciers melted, resulting in the sharp bends we see in them today and thus draining into the Esk. The small stream in the present day Mosedale is sent, however, to the east of Hard Knott, joining the River Duddon at Cockley Beck.

There is no lake in Eskdale in its more central section as there is in many of the other Lakeland glacial valleys; however, there have been lakes here in the past. River terraces in the area around Brotherilkild are former lake beds, now being cut through by the River Esk.

Another lake did exist during late stages of the ice age lower down the valley. Water was ponded between ice-free ground to the east (Birkby Fell and Corney Fell) and two separate glaciers. Ice flowed from Eskdale (north-east) and the Irish Sea ice sheet encroached from the west. The lake formed north-west of Birkby Fell and extended north-west beyond Muncaster Fell, which was also ice free.

Before the ice started to melt, the Irish Sea ice sheet butted up against Birkby Fell/Corney Fell. Streams of water escaping from the lake ran along the edge of the ice, cutting channels (at different heights as the lake level lowered) into the fell side which are still evident today. As the ice melted, the lake drained away leaving these channels (and some more on either end of Muncaster Fell) as evidence of its existence.

Great Langdale is justifiably one of the most popular places in Lakeland. The views of the Langdale Pikes as you come up the valley and, farther on, the circle of fells at the head from the valley, are both classics. In my view, however, the best view of the mountains ranged around the head of Langdale is gained from the north-western reaches of the summit plateau of Lingmoor Fell, on the south-western side of Great Langdale (the northern side of Little Langdale) (see Walk 13 and Photo w13.5).

Photo 6.12

Eskdale.

Although none of the glacial valleys we have looked at so far is entirely straight, none is anything like as convoluted as the connected valleys of Great Langdale and Little Langdale (although the Ullswater valley also turns sharply in several places as we will see). From its head beneath Rosset Pike, Mickleden (as this section of the valley is known) runs south-east as far as present day Stool End farm, where the glacier that carved the valley was joined by ice from present day Oxendale. Here the valley bends to the left, trending north-east and becomes Great Langdale.

There is also, at this point, a col separating Great Langdale from the valley containing Blea Tarn and Bleamoss Beck which becomes Little Langdale when it changes direction from south-south-east to east. There is a geological structural connection here as the great fault that runs from Ennerdale via Aaron Slack, Esk Hause and Rosset Gill to Mickleden continues through the col in the same south-easterly direction. It then carries straight on beyond the col, forming a deep wooded gully behind Bleatarn House on the south-western flank of Lingmoor Fell.

A subsidiary ice flow may well have crossed the col but the main flow carved its way along what is now Great Langdale, leaving the line of the major fault and running instead along a weakness created by earth movements during the mountain-building episode. The axis of an anticline runs through Great Langdale from north-east to south-west and continues up Oxendale. This anticline is the counterpart to the great Scafell syncline which is a key feature of Lakeland geology.

This feature determined the direction of the glacier flow for the next two kilometres, from Stool End to opposite the northern ridge of Lingmoor Fell. There, the glacier turned once again to resume a south-easterly direction to Elterwater and on to Skelwith Bridge and a rather more open landscape beyond the hard volcanic rocks of central Lakeland. This section of the valley has once again been determined by the presence of a fault.

East of Elterwater, there is a small lake impounded by tougher rock outcrops in the Skelwith Bridge area. There was also another lake, well to the west of Elterwater, beyond the narrowing area of the valley between Chapel Bridge and Elterwater. This lake has been filled in by sedimentation to form the present wide valley floor.

There is not very much to say about Thirlmere. The lake is now a reservoir and the original shores and lower slopes have been drowned beneath the high water level, hiding some features from view. The reservoir has been judiciously described as looking like a 'half-filled bath rather than anything nature could have devised'.

We should also note that Thirlmere lacks the classic U-shaped profile. There are low crags on the western flank, tough volcanic rocks made largely from lavas of the Lower Borrowdale Volcanic Group. To the east, on the other hand, the Helvellyn ridge curves gently from the summit area to steeper slopes lower down, but nowhere to crags characteristic of deeply incised U-shaped glacial valleys.

The fells on either side of the northern reaches of Ullswater are not generally considered to offer particularly good walking country.

Of High Seat, Wainwright said that walkers (as opposed to determined 'peak baggers') would 'justifiably consider its ascent a waste of precious time and energy when so many more rewarding climbs are available'. Equally, the rounded grassy slopes on the western flanks of the Helvellyn ridge are nowhere near as attractive as the sharply incised glacial features on its eastern side, including the famous Striding Edge and the run of glacial corries from Dollywaggon Pike to Whiteside. Given the choice of climbing Helvellyn from the east or the west, the answer is usually obvious.

This all means that Thirlmere is nowadays little more than a route to somewhere else. Cars, buses and lorries thunder, almost unaware of it, across the gentle pass of Dunmail Raise heading in search of the more glorious scenery around Ambleside and Keswick. However, seen from High Rigg for example, the Thirlmere valley is outstandingly beautiful. As Wainwright pointed out, while 'the greater part of the central [fells] ridge is marshy and the summit area dreary...Bleaberry Fell is a magnificent exception'. The ascents are fairly straightforward, the ground generally dry and the views exceptionally good.

We can now move across to have a quick look at the valleys of the eastern fells. The overall drainage pattern here is rather distinct from the spoke pattern we have so far studied. Indeed, in this area there is almost a repeat in miniature of the main, with valleys radiating from the High Street and Grey Crag area north, east and south. The eastern flank of the Helvellyn range add to the flow of drainage water into the major valley of the eastern fells, Ullswater.

Ullswater is the only large lake in the Lake District which has sharp bends in its course. The valley is also rather different in that a comparatively large number of valleys fed it with ice. Dovedale, Deepdale, Grisedale, Glenridding, Threshthwaite Glen and the Hayeswater valley all converge on the main valley within a stretch of about 5km.

The valley would also have been fed in the later stages of the ice age by a complex set of corries on the Hellvellyn and High Street massifs. Four corries on Helvellyn and six on High Street may well have supplied ice into the valley either around the same area and, in the case of some of the High Street area corries, to the east of Hallin Fell.

The upper reaches of the valley, as well as the southern and middle sections of Ullswater itself, follow the line of a set of complex faults. South of the lake these faults also mark the eastern limit of a caldera created during the Upper Borrowdale Volcanic Group cycle. However, faults alone did not determine the shaping of the valley.

The two great turning points in the trend of the lake, between Silver Hill and Glencoyne and between Geordie's Crag and Skelly Nab, are both marked by rock bars which cut across the lake. The lake is over 40m deep where it is squeezed between Stybarrow Crag and Devil's Chimney, but it becomes shallower as it approaches Norfolk Island (between Silver Hill and Glencoyne). Indeed, Norfolk Island is part of one of the rock bars which rises up here to break through the level of the lake. These rock bars may (or may not) represent pre-glacial cols, but what is clear is that the rocks on either side have presented serious obstacles, causing the ice to find a less resistant route than straight ahead.

An 'outlier' of sedimentary rock from the Skiddaw Group (known as the Tarnmoor Formation) has been thrust up here, on fault lines, to break the surface between the outcrops of the Lower BVG lavas that otherwise dominate the fells north of Glenridding. More of the same rocks are found farther north and to the north-east, along the line of Ullswater on the lower flanks of Barton Fell.

The southern limit of these Skiddaw Group rocks forms the northern shore side of Ullswater from Glencoyne to Aira Point. This is where the lake is deepest. For a stretch of about 1km, south-west of Aira Point and Low Birk Fell, it is over 60m deep, rising to about 20m by Skelly Nab and Geordie's Crag where another fault displaces the Skiddaw Group rocks across the lake. The fault line follows the north-eastern face of Hallin Fell. Immediately past this constriction the lake becomes somewhat deeper again, reaching over 30m, then gently becomes shallower as it flows to its end point at Pooley Bridge.

The ice carving out the Ullswater valley has exploited the weakness left by these faults and associated thrust movements, turning where necessary to avoid tougher rock in its direct line of advance. It takes an effort of the imagination to conceive what the glacier would have looked like from above as it flowed, widened, narrowed, deepened, twisted, widened and deepened yet again, until at last the ice was free of the pressure of resistant rock and could spread out. A good place to try and envisage the valley in the ice age can perhaps be found from many of the minor peaks on either side of the lake, such as Hallin Fell or Sheffield Pike, but you will need to contribute a good deal of imagination.

Once, Ullswater would have reached much farther back along the valley, to Brothers Water. Sediment carried down by streams from the surrounding fells has been deposited here, filling in the lake. After heavy rain, some parts of the land just south of the lake are flooded and access to the campsite has to be via the tarmac road to Rooking. Eventually the fells will be reduced, year by year over millions of years, until they are much lower than they are today and the lake will be entirely filled in.

The side valleys running into the main valley are themselves superbly scenic, both from within and when looking down on them from above. Dovedale, Deepdale, Grisedale, Glenridding, Threshthwaite Glen, Hayeswater, Boredale, Bannerdale and Ramps Gill are all immensely beautiful and provide dramatic evidence of glaciation in their steep-sided valleys and crags, as well as in the serried ranks of moraines that often litter the valley floors. A good walk on

Photo 6.14 | Brotherswater (left) and Deepdale (right).

a cloudy day – or even a wet one – can be had by walking up and down one of these valleys, especially Grisedale where a walk from Patterdale or Glenridding to Grisedale Tarn and back (much of it on two different paths) is a delight.

It is interesting to note that these valleys flow mainly north or north-east into the main valley which, as we have seen, itself heads north and north-east.

To the east of the High Street ridge (from end to end the longest continuous ridge in Lakeland) another set of minor valleys flows east (photo 6.14). The best known is Haweswater, another lake turned into a reservoir. To the south, yet more minor valleys flow from the Helvellyn and High Street ranges due south (Rydal Beck, Scandale, Stock Gill, Troutbeck, Kentmere and Longsleddale). With the exception of Kentmere, these are narrow but long valleys, running between long narrow fingers of the fells.

The narrow sides mean that walking in these valleys is a confined experience rather than the expansive feeling from walking atop the intervening ridges. As a result, they are pretty quiet places for exploration (especially Rydal Beck and Scandale where the marvellous Fairfield horseshoe (Walk 10) is usually a stronger attraction). However, they do provide a different view of the scenery and can usefully be combined with an ascent to a summit via the valley head. Walking in these valleys you will certainly come to appreciate the size and form of moraines (see photo w10.5).

Photo 6.15
Eskdale from
Border End. Left
to right: Scafell,
Scafell Pike,
Great End,
Esk Pike and
Bowfell.

To the east of Ambleside, however, in Stock Gill, Trout Beck, Kentmere and Longsleddale, the valleys all become noticeably wider. This marks the boundary between harder volcanic and softer sedimentary rocks. The glacier flowing out of each valley suddenly found the rocks less resistant and so was able to carve out a wider channel.

We have now completed a rather long tour of the major glacial valleys of Lakeland. I think it has been worth looking at the valleys in some detail as they are an important feature of the scenery. Indeed, for many visitors, they are the key to the enchantment and beauty of the area. Although the fells are undoubtedly the main draw to the area for hillwalkers, it would be churlish for even the most dedicated peak-bagger to deny the glory of the interplay of valley and fell to Lakeland's landscape.

In any case, the valleys are necessary to get from one place to another in the area and therefore everyone has to follow the flow of ice and then water out from the centre of Lakeland to its periphery. There are plenty of opportunities to study what these glacial trenches look like from within, now that the ice has gone. They are most interesting when on foot rather than on wheels when more of the subtle detail and the atmosphere can be seen and felt. Anyway, some of the loveliest of the valleys of Lakeland are only accessible by pedestrian power and should not be ignored.

Before leaving the ice age, we need to look at one more characteristic feature of glacial gouging: the glacial carving of corries. This will be the topic of the next chapter.

CHAPTER 7

Corrie Carving

Corries are an utterly enchanting aspect of the glacial landscape. In classical form (with a deep rounded bowl, a rock bar at the threshold and an impounded circular lake, suspended half-way up a mountain), they are undeniably the most enthralling and often most atmospheric feature of the mountainscape. Looking down on a corrie from above, its glacial origin is almost tangible and it is easy to imagine the bowl filled with a river of ice that flowed over the corrie lip or threshold. Looking up from within even a modest corrie, the sheer size of even a small glacier is overwhelmingly obvious. Stickle Tarn, Low Water, Blea Water and Red Tarn are just a few classic glacial corrie lakes and all are justly adored by fellwalkers for their stark beauty.

Another glacial feature of the landscape (one which often provides for exhilarating hillwalking) is intimately associated with corries, namely the 'arête'. This is a sharp, narrow ridge and is the product of the glaciers in neighbouring corries or valleys, each cutting back into the mountain leaving only a narrow wall of rock between them. Some of these resulting

Photo 7.1 | Low Water, Coniston Old Man, a classic glacial corrie suspended halfway up a mountainside in a north-east facing bowl.

arêtes are prime among Lakeland's most scenic wonders, as well as providing some of its most sensational walks. Striding Edge on Helvellyn and Sharp Edge on Blencathra, part of the Skiddaw massif, are very fine examples. In this chapter we will complete our review of the glacial landscapes of Lakeland with a look at some of the most popular corries, corrie tarns and arêtes of the area.

Not all corries are deeply enclosed bowls, not all have rock bars at their threshold, not all contain tarns and not all have neighbours close enough to result in the repeated shaving of the rock on two sides of a ridge needed to create an arête. As ever in geology, there is a wide variety of forms and in extreme cases it can be difficult to decide whether a bowl on a mountain is a corrie, an 'immature corrie' (i.e. one which is in the early stages of development and sometimes known as a 'nivation cirque') or just a dip in the land.

Different studies by geomorphologists have identified between 50 and over 150 glacial corries in Lakeland. Most of them are facing north-east, although there are a few east to south facing corries and a few north to north-west facing ones. The bulk of them, however, look firmly to the north-east, carved high into mountainsides below the summits.

Two factors favour this pattern of development of corries. First, the north-eastern slopes are sheltered from the prevailing winds so snow can collect without being blown away. Sec-

Photo 7.2 │ A cluster of glacial corries. From left to right: Cock Cove, Ruthwaite Cove, Nethermost Cove, the ridge beyond Nethermost Cove with the summit of Helvellyn just peeking over the Edge to the right (seen from St Sunday Crag). Arêtes separate the corries, with Striding Edge forming the ridge at the top right of the photograph.

Photo 7.3 | Nethermost Cove, Helvellyn Range, one of the clusters of glacial corries between Whiteside and Dollywaggon Pike. Nethermost Cove no longer has a corrie tarn as it has dried up due to the lack of a rock bar to hold it in.

ond, the north-easterly aspect results in a minimum of sunlight so the snow that collects does so without being melted away so easily.

The snow collects in a dip or hollow or perhaps an existing gully. As it accumulates over several years, the weight of the snow increases and presses down ever more heavily on the lower layers of snow turning them progressively into ever more dense ice. Eventually the weight of the overlying ice and snow causes the ice at the lowest level to deform and move outward from the slope. As it does so it cuts down into the mountainside, deepening the original dip or hollow. The glacier undercuts the sides and the back of the slope creating steep corrie walls if the rock is resistant.

If the underlying geology is suitable the corrie glacier may even cut out a deep bottom, lower than the height of the threshold, resulting in a rock bar. When the glacier eventually melts the presence of a rock bar may then allow water to accumulate to form a lake. If there is a rock bar then it means that the ice had to move upwards over the rock bar to escape from the corrie before it could start to flow down the mountainside. This upward movement illustrates the immense forces created by the weight of accumulated ice and snow.

However, it should be pointed out that a lake can also be impounded by moraines, so not all glacial corrie tarns are retained by a rock bar. Also, as mentioned above, many corries do not have a glacial tarn. This is because the glacier will choose the path of least resistance and if it is formed on top of a geological weakness, such a softer layer of rock between two layers of harder rock, no rock bar is formed.

Blea Water on the eastern side of High Street is a classic example of a glacial corrie, complete with rock bar, tarn and arêtes on either side, separating it from Riggindale to the

Photo 7.4 |

Corrie glacier tumbling out of the corrie as an icefall (Tour de Mont Blanc). Photo Reg Atherton.

Diagram 7.1 |

Formation of corrie glaciers.

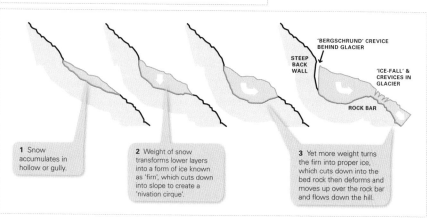

1 Snow accumulates in hollow or gully.

2 Weight of snow transforms lower layers into a form of ice known as 'firn', which cuts down into slope to create a 'nivation cirque'.

3 Yet more weight turns the firn into proper ice, which cuts down into the bed rock then deforms and moves up over the rock bar and flows down the hill.

'BERGSCHRUND' CREVICE BEHIND GLACIER

STEEP BACK WALL

'ICE-FALL' & CREVICES IN GLACIER

ROCK BAR

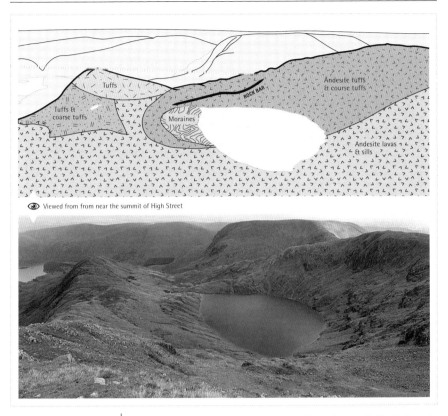

Diagram 7.2 + Photo 7.5 | Blea Water (High Street range), a glacial corrie tarn impounded behind a rock bar of comparatively harder rocks than those underlying the area of the tarn. Blea Water exists because the mixed geology of this area meant the glacier dug down into the bed rock before it could escape over a 'rock bar', while neighbouring Riggindale does not have a glacial tarn as it was cut out along the line of a fault.

north and Small Water to the south. The tarn is currently over 60m deep and the lowest ice had to rise nearly 100m to get over the top of the rock bar (which also has a thin layer of moraine on top of it).

Interestingly, neighbouring Riggindale was also a glacial corrie, but it lacks the characteristic bowl, rock bar and tarn. The underlying geology explains the difference between the two adjacent corries. Blea Water is impounded within an area of pyroclastic rocks which run across the corrie, so that the lake is situated on softer rocks while comparatively harder rocks form the rock bar. Riggindale, however, lies along the line of a fault in Upper Borrowdale

Volcanic Group lavas. A boundary between the lavas and outcrops of pyroclastic rocks runs just south of the fault on the valley side, parallel to the line of the fault.

These geological features have provided a weakness that the glacier could exploit without being forced to carve down into the floor of the corrie bowl before the ice could escape. A tiny bowl, called Sale Pot, sits in the north-western corner of Riggindale at a height of about 525m and represents the final corrie seat of the glacier that once existed here at the very end of the ice age.

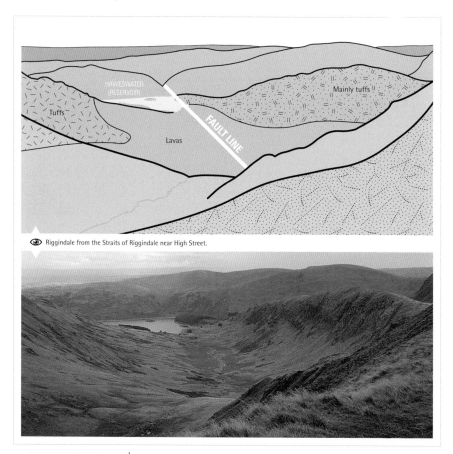

⊙ Riggindale from the Straits of Riggindale near High Street.

Diagram 7.3 + Photo 7.6 | Riggindale, a glacial corrie lacking a corrie tarn. A fault runs along the line of the stream and a geological boundary runs along the right-hand cliffs, Rough Crag, illustrating that the glacial valley has developed along the line of a fault.

Photo 7.7 | Angle Tarn, a glacial corrie sheltered below the east-facing crags of Esk Pike and Bowfell.

Lakeland's corries are generally grouped in small clusters e.g. around Coniston Old Man, High Street, Helvellyn, the Skiddaw massif, Grasmoor and south-west of Buttermere. A much looser cluster of corries also exists, forming a wide arc in the central fells ranging from Great Gable, north-east of the Great End/Esk Pike range, Langdale and east of High Raise.

Angle Tarn, below the col between Esk Pike and Bowfell, is a classic corrie lake, almost circular in shape and impounded behind rock bar. It is oriented north-east and sits in a very well-sheltered spot.

Another cluster of four more corries can be seen on the south side of Coniston Old Man and another two on its north-eastern side. A walk over Dow Crag, Coniston Old Man, Swirl How and Black Sails on Wetherlam is an ideal way to view these beautiful but bleak corries, especially if combined with an ascent and descent on the tracks that trace through the corries themselves. All four tarns are effectively impounded due to the underlying geology of tough pyroclastic rocks forming rock bars. Levers Water and Goat's Water both face onto rhyolitic welded tuffs, while Blind Tarn and Low Water butt onto coarse dacitic tuffs.

Low Water (see Photo 7.1), Goat's Water and Blind Tarn (Photo 7.8) are all especially atmospheric. Blind Tarn (only really visible from the ridge between Dow Crag and Brown Pike) is particularly appealing as it is perched improbably on a ledge on the southern end of the range. It is tucked away behind a small easterly projection in the final part of the ridge (the site of quarry workings). Blind Tarn's apparently precarious and unfavourable position shows how even small factors can influence whether sufficient snow and ice can collect in a dip or hollow for a glacier to form.

Photo 7.8 | Blind Tarn, perched improbably on a small ledge on the southern end of Dow Crag, below Brown Pike.

More corries and glacial tarns can be seen in the Langdale Pikes area, especially the highly popular Stickle Tarn below Harrison Stickle and Pavey Ark. Stickle Tarn is also impounded behind a layer of tough volcanic rock.

A fascinating feature is the nearby Langdale Combe, located a couple of kilometres to the north-west of Stickle Tarn. This is a glacial corrie according to some authorities, but the most compelling aspect is the mass of hummocky moraines left by other glaciers that passed through here (see Photo 6.8).

Photo 7.9 | Stickle Tarn with Pavey Ark crags on the left. Note the hummocky terrain behind and to the right of the tarn and the rock bar impounding the lake (with some hummocky moraines on top of it).

Photo 7.10 | Red Tarn (left) and Striding Edge (right).

The cluster of corries on the eastern flanks of the Helvellyn range is particularly impressive. One of the best views of the cluster is to be found on St Sunday Crag, looking towards Nethermost Cove, Ruthwaite Cove and Cock Cove (see photos 7.2 and 7.3). Striding Edge rises high above the smaller arêtes seen between these corries and hides Red Tarn from view (photo 7.10).

Another corrie stands to the left of Cock Cove, Grisedale Tarn (photo 7.11). The western side of the Helvellyn range consists of sloping grass, with few indentations let alone outcrops. However, the eastern side of the same ridge is a riot of rock, with a run of corries suspended high up below the summit line. While every year a few thousand climb Hevellyn and its neighbouring peaks from the west, several hundred thousand toil up the steep rocky ridges to the east, attracted by fell scenery par excellence.

The area around Grisedale Tarn is most unusual. Rather than a corrie bowl with high side walls and back wall, Grisedale Tarn sits in a col between Fairfield and Dollywaggon Pike with a low rounded hump (Seat Sandal as its backdrop). Steep valleys fall away to either side of Seat Sandal picking a route between the crags, Reggle Knott to the east and Tongue Gill to the south. The Tarn sits in a bowl carved improbably out of the surrounding rock and is a most intriguing sight, both from afar (best from the ridge between Fairfield and St Sunday

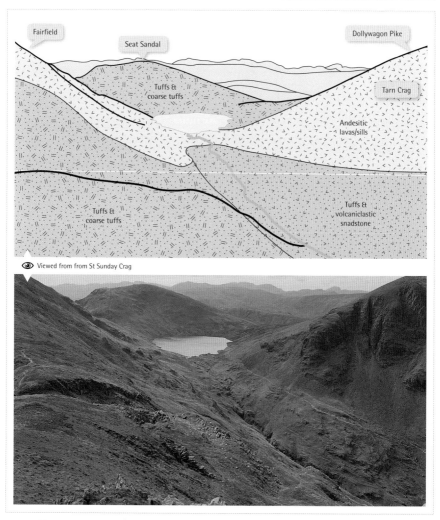

Diagram 7.4 + Photo 7.11 | Grisedale Tarn, another corrie tarn impounded in an unlikely site beyond the high side walls formed by the crags rising to Dollywaggon Pike (right) and Fairfield (left).

Crag) and from within. The summit of Seat Sandal is made up of hard andesite lavas, as is the area immediately north, east and south of the tarn itself. Comparatively softer rocks make up the cols to the west and south-west which have been eroded away, but not enough to allow the water to drain west or south-west. Instead it heads east through the deep glacial valley of Grisedale fed by the corrie glaciers high up on the ridge. The lake has been impounded by this interplay of rock and mountain ridge.

Grisedale Tarn and Grisedale valley are on a highly popular path in its own right (the 'Coast to Coast' Walk) as well as having popular peaks to the north (Dollywaggon Pike and Helvellyn among them) and south (Fairfield and its long south-pointing ridges). This means that this area is always very busy, but by going up into the upper corries you can often manage to leave the crowds behind. This is rough going; there are some steep slopes and the need for a bit of scrambling at times if you hope to reach the ridge line, so may not suit all walkers. Also, a walk up Grisedale to near Ruthwaite Lodge and then into one of the corries and back down the valley on a different path makes for a good walk on a day when the summits are sheathed in cloud.

For good clear days when the summits beckon, Walk 8 includes an overview of these corries and also passes along a route with stunning views of Red Tarn, cupped in a classic corrie bowl beneath Helvellyn and bounded to the north and south by two great arêtes, Striding Edge and Swirral Edge. Part of the route crosses Striding Edge where the walker can peer down into great corries both to the left and to the right, one impounding a tarn.

The Red Tarn corrie has been cut into slightly softer rock between the harder arms of rock that make up Striding Edge and Swirral Edge. However, Nethermost Cove and Ruthwaite Cove have both been carved into faults in the same rock as the Red Tarn corrie. The last of the series of corries here, Cock Cove, has exploited the weakness created by the boundary between this pyroclastic rock and the andesitic lavas, mentioned above, that impound Grisedale Tarn.

In this area the effect of glacial cutting of corries, arêtes and deep valleys has been to produce an utterly fascinating mountain landscape, one that displays the full complexity of the interplay of rock type, faults, mountain-building and glaciation in determining the landscape we see.

The same is true of Walk 9 which covers a vast variety of glacial features on and around High Street, another area of intense corrie development. We have already mentioned Blea Water (see photo 7.5), impounded in a basin behind a rock bar although the rock bar itself is covered by a smattering of moraine. The route also goes along the rather wide arête Gray Crag between two glacial valleys, Hayeswater and Threshthwaite Glen. Both these valleys had corrie glaciers in them, although neither is a classical bowl shape. Instead, the glaciers sat in shallow areas near the head of the valleys. Although they cut into the valley sides creating the deep valleys we see today, they did not create a sharp arête. The walk along the blunt ridge is probably on the pre-glacial surface, as is the section of the walk along the summit of High Street itself.

More fine corries can be found on the Skiddaw massif, especially on the southern flank of Skiddaw. For some reason, ice could gather here despite the south-facing aspect (although there are also east, north and north-west facing corries on the massif). The height of the area must have been important in sustaining these southern corries which have produced some fine arêtes, although the corries themselves are not particularly classic in form.

The best corrie on the Blencathra faces east and has a lake, Scales Tarn. It also offers the famous ridge walk on the appropriately named arête of Sharp Edge. This is a tougher challenge than Striding Edge and a good head for heights is essential if you intend to tackle this exposed scramble (on which fatal accidents do occur, so recognise your own limits before undertaking such a climb). An alternative that avoids the scramble is to take the track from within the corrie that heads south of Scales Tarn up to the summit of Blencathra. This allows you to see the corrie from a variety of viewpoints and thus to appreciate its fascinating shape bequeathed to us by the retreating ice.

Farther west there is a fine set of corries on the High Stile ridge on the south-western side of Buttermere. A tarn is impounded in the corrie below High Stile and Red Pike, but not in the other corries here. A walk up through Birkness Comb, via a steep scramble onto the northern ridge of High Crag and then over High Stile and Red Pike, gives an excellent view over these corries.

If you continue and descend via Scale Gill, you will drop down into a delightful, classic V-shaped stream-cut valley, eventually coming out somewhere above the top of Scale Force.

Photo 7.12

A blunt glacial arête, between Thornthwaite Crag and Gray Crag.

Photos 7.13 (top) + 7.14 (right)
Scale Beck on the north-western flank
of Red Pike drains through a V-shaped
stream valley. It reaches the over-
deepened lower valley draining Black
Beck high on the valley side, so the water
suddenly plunges from the 'hanging
valley' in a fine waterfall, Scale Force.

Unless you know the lie of the land, you will have to drop down to the bottom of the waterfall to get a view of it from below (see photos 7.13 and 7.14). This is an example of a 'hanging valley', where the stream from above previously drained into the valley at the top of the waterfall. Glaciers in the lower valley deepened it more than the subsidiary valley, so that when the ice retreated the stream's upper valley was left 'hanging' and a waterfall developed.

You can see many such hanging valleys and waterfalls in Lakeland. Indeed, the corries around the corner from here are also hanging valleys and drain via waterfalls, such as Sour Milk Gill, which you can pass on the way back to Buttermere. The waterfalls that occur all around Lakeland are the signs of hanging valleys: streams left stranded by glaciation.

Some corrie glaciers may well have been the very first glacial features to have developed on the upper reaches of the fells at the beginning of the ice age, 1.5 million years ago. They may have existed for thousands of years, going out of use as the climate warmed up and coming back into operation as it cooled down once again. At times, the corries were taken over by the much more massive phenomena of the coldest periods of the ice age: ice caps and valley glaciers. As conditions once again became warmer the ice cap and the valley glaciers retreated, eventually to melt fully. Once again it was only in corries, at the higher levels, that enough snow could still accumulate and become sufficient to create glacier ice.

All the corrie glaciers started to melt about 15,000 years ago, those of the Helvellyn and High Street ranges lasting longest. However, there was a sting in the tail to come. About 10,500 years ago there was a short resurgence of cold conditions and once again glaciers developed in the upper corries. This resurgence, known as the 'Loch Lomond Advance', did not last long. The climate has fluctuated somewhat since then, but has been more or less steady.

When the ice finally melted, it left the crags forming many corrie back and sidewalls with over-steepened slopes. As a result, there were many rockfalls as rocks higher up peeled away from the main body of rock. These fallen rocks accumulated at the bottom of the crags to form today's scree slopes. It is therefore necessary, just as with the deep glacial valleys, to try to imagine the corrie walls without the scree to get an impression of the true size of the glacier that once occupied the corrie.

By about 10,000 years ago all the corries became devoid of ice, some of those on High Street being the very last places to become ice free. Many more of the corries would have had a tarn than is the case now, but in many corries any such tarn has now been filled in by sediment, often leaving a boggy corrie floor. However, many tarns do still exist and today contribute greatly to the scenic charm of the Lake District, as do the arêtes, deep valleys and summit areas left to us after the ice age had done its work on the pre-glacial landscape. However no sooner had the ice retreated than many new influences on the landscape started to appear on the scene.

CHAPTER 8

Toolmakers to Tourists

At the end of the ice age, the ice retreated leaving behind bare rock. It was not long before flora and fauna moved in, including human beings. Since then developments have been momentous and the pace of change is steadily increasing due to human influence.

The ice had scraped away all the vegetation that existed before its arrival. At the end of the ice age, the fells were therefore entirely rocky. At first the temperature was still fairly cold so that freeze-thaw action was particularly effective in breaking up the exposed rock. The glaciers also left behind unstable slopes in valleys and corries. Gravity, aided by freeze-thaw, pulled the unstable rocks away from the over-steepened crags to create the vast screes which today curtain the lower slopes of valleys and corries.

Lakes collected wherever there was a block to free drainage e.g. in corries, in dips and hollows scrapped out by the ice on the fells and in the deep ex-glacial valleys.

Rock which is exposed to the atmosphere is attacked chemically in a process known as 'weathering', whereby chemical changes occur within the rock contributing to its breakdown.

The first living coloniser of the rock was lichen, a combination of fungus and algae. Its reproductive spores are carried in the air and it was soon spreading across the rocks, extracting minerals from the rock and speeding up the process of weathering and the creation of soil. There are many thousands of species of lichen, some preferring specific types of rock.

Lichen of different types often completely coats the outer surface of a rock, sometimes making it hard to see what type of rock you are looking at. From a distance a rock may look like a volcanic breccia or coarse tuff, but on closer inspection it turns out that what look like lumps are in fact lichen (see photo w7.2). However, these are only the visible lichen. There are many more species of 'micro-lichen' which can't be seen by the naked eye but can be distinguished when using a hand lens. These micro-lichens are said to be responsible for making rock slippery when it is wet.

Lichens start the process of making soil and once soil is present plants can begin to literally take root. Grasses, mosses, ferns and some dwarf artic plants would have been among the first vegetation in the area. This was closely followed by 'pioneer' trees such as juniper and then birch and pine. Hazel trees added to the variety of early birch woodland from about 9,000 years ago, then oak and elm also took root.

By about 7,000 years ago, woodland probably covered almost the entire area below about 650m (although the actual 'treeline' is lower in the wet and windy western fells than on the more sheltered eastern fells). Almost as soon as they had established their presence, the trees came under increasingly sustained attack. There is evidence of systematic tree clearances on the uplands by the early human farming communities from about 5,000 years ago.

The axes used to fell the trees were made locally from flinty volcanic rocks. Small groups of people, using such axes and aided by fire, could clear significant areas of trees quite quickly. As tree cover declined, so grass, herbs and weeds became more prevalent. Initially the cleared areas were used for animal pasture rather than for growing cereals. When an area was over-grazed it was temporarily abandoned to be taken over by bracken and, if left untouched for long enough, perhaps by regenerated woodland.

The rate of tree clearance increased substantially about 4,000 years ago and shortly after that cereal production started to become important. The evidence suggests that, except for the valleys, upland tree clearance was effectively complete and few if any original areas of upland woodland exist today.

Some time after 3,000 years ago, the climate became wetter. Bog developed as a result of leaching and soil compaction by herds of grazing animals on flat areas which had been cleared. Fewer than 2,000 years ago, Norse farmers also began to clear the lowlands of trees, and the outlines of today's Lakeland landscape was essentially complete. Some woodlands may have survived to become coppiced woods for the production of charcoal on lands of Cistercian abbeys (see below), although the modern plantations that cover areas such as Ennerdale are a later contribution.

This underlines the point that human beings do not only remove things, but also alter and add to them. Wood and stone were the tools of early human beings in the area. These tools cleared forests, but also helped build settlements.

I mentioned above that local flinty volcanic rock was used to make axes. Several sites where axes were hacked out of the rock and then chipped into practicable shape have been identified in Great Langdale, on Scafell and elsewhere. These axes were not only used in the immediate locality. Axes that can be identified as being made from Lakeland rock have been found in many different places in the British Isles. The axe trade must have become quite important. It has become standard to describe the Langdale and other sites as 'axe factories'.

Photo 8.1 Pike o'Stickle and the top of the gully which was the source of material used to make neolithic hand axes.

Several sites have been found: at the top of a scree gully next to Pike o'Stickle; on the southern face of Harrison Stickle; near Mart Crag; Glaramara; Great End; and close to the summit of Scafell Pike. Archaeologists have also found small piles of rock chipped off to shape rough axes near Pike o'Stickle and Mickleden: the first waste tips of those who exploited Lakeland's rock resources. Archaeologists believe that only 'rough chipping' was performed here with the finer finishing work being done elsewhere, the rough sets being transported across the earliest regular human tracks on the fells.

The sites were almost certainly not occupied all the year round and were only in use in spring and summer. It seems that settlements were confined to the lower fells, where the flinty volcanic rock is not present.

What is clear is that neolithic humans must have clambered across and studied the high fells in very close detail to search for the type of rock which they could exploit. We are only the latest in a long line of fellwalkers stretching back for some 5,000 years. Interestingly, one chipped flinty axe-stone was found on the main path from Wastwater via Lingmell Gill to the summit of Scafell Pike (covered in scratch marks from recent generations of booted fellwalkers).

It is probably the case that we may even literally follow in their footsteps, using many of the same tracks as neolithic fellwanderers. Indeed, archaeologists believe some major routes were probably in use from about 4,000 years ago, such as from the head of Langdale, via Rosset Gill, Angle Tarn, Esk Hause, Styhead Pass and Aaron Slack to Ennerdale and beyond, and the route from near present day Windermere across the High Street ridge to the north.

Other pre-historic remains include cairns, stone circles and ruins of buildings, especially circular houses. Many cairns, of which there are over several hundred and which mark burial places, are found on outer fells, such as Corney Moor in the far south-west of the national park, Stockdale Moor straddling the park's western boundary, around Ullswater and around Shap Fell to the east (although there is also a smattering of cairns on the north-central fells south of Thirlmere and Ullswater). Like the cairns, circular stone settlements are found mainly on the lower fells and there is none to be seen when walking the high fells.

Several stone circles are found to the north of the central fells area, running in an arc from Elva Plain in the north-west via Castlerigg and Motherby, to Shap Stones and Gamelands in the east. More such circles can be found to the west and south-west, making a total of seventeen such monuments to stone culture.

Castlerigg is the stone circle that receives the most visitors. Indeed, some of the country lanes near it have been stretched to motorway width to accommodate the coach parties. It is of course best approached on foot, just as it would have been in its day, by integrating it into a wider walk (see Walk 12). This allows the scenery, and thus the setting of the stone circle, to be appreciated. Archaeologists have, in tandem with the growth of modern interest in 'landscape' studies, recently begun to place great emphasis on the role of the environmental surrounds of prehistoric monuments in understanding them.

Castlerigg, with its thirty-nine stones, sits atop a low rounded eminence with wide views over the surrounding rural and fellside landscapes (depending on the extent of tree cover that

Photo 8.2

Castlerigg

stone circle.

had been removed from the area when it was constructed). The stones are made of volcanic rock, both lavas and tuffs, and were either transported here by humans or by a fortuitous dumping here by glaciers of such an eminently suitable set of stones.

The building of stone circles was common to the neolithic communities of the western seaboard of Europe from Spain, France, Brittany, Ireland, Wales and Scotland. This suggests that ideas must have been traded as well as axes. Much more than this we cannot really say, however much archaeologists may speculate.

In the 1970s, when the electronic computer and calculator were making an impact on wider consciousness, it became commonplace to think of such stone monuments as astronomical calculators and observatories. Increasingly aware as we are of the fragility of human life and indeed planetary life, this is a now a quaintly anachronistic concept. No doubt our present speculations will also prove to be out-of-date when viewed in the light of future thinking.

What is not clear is whether we will ever be sensibly able to recreate the neolithic mind-set. What we can do, however, is to soak up the surrounds and the scenery and consider its place in the circle of standing volcanic stones.

Whatever the role of stone circles, it was only a temporary one and within a few hundred years they became abandoned signs of a decaying culture. The bronze age introduced metal to the toolset of human beings, opening new opportunities to relate with the environment and develop human societies. About 1,000 BC (i.e. 3,000 years ago) there was a sudden expansion of farming land to higher sites, aided by a period of higher temperatures. Soon iron also became a tool of the inhabitants of the area.

It is to these early human activities that we owe the open, rugged landscape that we come to marvel at as showing 'natural beauty' or 'natural wilderness', or more inaccurately still as 'pristine' landscapes. Following the felling of the natural woodlands, increased soil erosion led to the creation of deep gullies on the fellsides, to a speeding up of the rate at which lakes were filled in by sediment and to the creation of boggy land.

Apart from Carrock Fell there are no high-level iron age 'hill forts' in the Lake District though some ramparts and ditches in the main Lakeland valleys may well be of iron age origin, perhaps from the time when the Roman empire was first reaching out to encompass southern Britain within its grip.

One reminder of Roman times is the track that crosses the High Street ridge from near Windermere to near Penrith. Its rather oddly urban-sounding name derives from being a high-level track rather than from being the main thoroughfare of a settlement.

Photo 8.3
High Street.

The Romans have also left us the traces of some of their forts, the best example being Hardknott Castle in Eskdale, or Mediobogdum as it was known to them. Another Roman fort was set up at Ambleside, linked to Brougham fort near Penrith, by the track over High Street.

It used to be assumed that the ring of Roman forts and stations around the fells was aimed at controlling the wild inhabitants of the uplands, but this idea has now gone out of fashion and it is believed that the fells were only lightly populated and thus needed little control. The forts and stations were integrated into the complex set of installations that provided rear security to the symbolic boundary of Hadrian's Wall.

It seems probable that the Roman presence in the area was primarily aimed at securing the north-western marches or wide borderlands of the empire in Britain against incursions from Scotland and Ireland. Since their interest was in the lowlands of England, and their potential for providing wheat to feed the empire's population, it is unlikely that inclusion of the Lakeland area in the economy of the empire was the aim. There is no evidence, for example, that the Romans made any effort to exploit the mineral resources of the Lake District. All in all, compared with other parts of Britain, the Romans had little effect on the landscape as we see it today.

In the post-Roman era (what used to be called the 'Dark Ages' but is now more commonly referred to as the early medieval period) different waves of settlers arrived from northern Europe and made inroads into the area, adding to the rate of tree felling in lower lying lands and the expansion of agriculture. Little or no written evidence exists from this time until the central medieval period, when the Anglo-Norman kingship expanded its reach into the north-west at the end of the 11th century AD. The only evidence from the early medieval period is archaeological or related to place names.

Blencathra is an old British name (from the language that developed into modern Welsh) as are some river names e.g. the Cocker, Derwent and Esk, but many modern place names derive from Norse, Anglican and Viking settlers. Mere (lake), thwaite (clearing) and garth (yard/garden) are all common names derived from northern European languages. For example, Stonethwaite means 'stone clearing'.

From the 9th century AD settlers began to move into the deep valleys around Scafell, Great Gable, Helvellyn and High Street, removing trees and shifting stones. This was a process that went on for 300–500 years, well into Norman and Angevin times. It was these Scandinavian settlers who started the human habitation of the heart of Lakeland, albeit in the deep valleys.

In the Norman era came economic expansion in the form of new market towns, fairs and the foundation of monasteries, often bringers of new forms of agricultural techniques, methods and related economic activities. The most important novelty of these ages for the fells was the introduction of intensive sheep rearing (although sheep were probably first introduced to the area back in the neolithic age).

The abbey at Shap was the only one to occupy a site in the uplands, but even those in the lower lying areas were active in promoting sheep grazing of the fells. The modern place name element 'grange' means a remote sheep farm belonging to a monastery. The monks have also left their mark in the now barely noticeable ditches and banks that marked the boundaries of their lands (such as in Eskdale near Throstle Garth). The number of sheep multiplied rapidly. Shap even gets a mention in a document of 1315 written by an Italian wool merchant, as one of his main sources of wool in Britain.

The sheep also stamped their mark on the fells environmentally, by increasing soil compaction and therefore rainwater run-off and by eliminating shrubs and other plants, enforcing a near monoculture on the fells. In other words, sheep turned the fells into the present-day landscape of farmed land. Brotherilkild in Eskdale was one of the first places to stock Herdwick sheep some time in the 10th century, and had become a farm with 14,000 acres of grazing land in the central medieval period belonging to Furness Abbey.

Large parts of the upper fells were declared to be 'forests', meaning the hunting ground of the local earls (forest as a place with a lot of trees is a modern meaning). Much of the Skiddaw massif and the area between Derwentwater and Bassenthwaite Lake and upper Eskdale were forests, as were Grisedale in the shadow of St Sunday Crag and the area around Haweswater. Legally, these areas could not be used for farming or settlement, being retained as 'wildernesses'. However, in the conflict between sheep and game, the economic power of wool slowly became dominant, aided by a relentless expansion of settlements. By the 16th century the forests had become agricultural land of one sort or another and settlements had expanded into the deepest recesses of the fells.

Many of the farms that exist in the valleys today were first developed in this central medieval period. The walls that delimited farm from upland pasture (the 'intake wall') also started to

Photo 8.4 (top) | Sheep have been pastured on the fells since the early medieval period.

Photo 8.5 (left) | Small lump of hematite (an iron ore), also known as kidney stone; the lump is about 2cm long.

appear though most of the stone walls we see today are of a later build. A walk up a major valley such as Mickleden or Wastwater reveals wide river beds strewn with hundreds of thousands of small rounded boulders (rounded from being battered against each other when the rivers are in flood). To the sides of the river bed, however, there are green fields separated by stone walls.

These boulders making up the stone walls are also rounded. It is clear that they were once spread across a much wider area than the present river bed and were painstakingly removed to make the stone walls. Great piles of boulders can even be seen near Wasdale Head, there simply having been too many to use them all in making stone walls.

However, there was an end to expansion with the arrival of the plague, war and starvation of the 14th century. Recovery did not really take off until the 16th century, with the dissolution of the monasteries and the sale of their lands to new owners along with the creation of a new class, yeoman farmers, who were well established by the 17th century. They built new farmhouses and outbuildings on the farm sites, creating many of the buildings we see today. Between about 1550 and 1800 the 'statesmen', as the yeoman farmer class became known, had successfully taken over and enclosed all the remaining open land.

The 16th century also marked the expansion of exploitation of the area's mineral wealth when the new landowners brought mining experts from Germany, then centre of Europe's burgeoning mining industries, to show them how to dig copper and lead out of the ground.

Mineral extraction and metal-working were not new to the area. Monasteries had often been centres of iron working with scattered 'bloomeries' such as that in Langstrathdale. The iron ore worked by the monks (or more accurately by their local workforce) was probably obtained from the col known as Ore Gap between Esk Pike and Bowfell. Here you can find tiny lumps of iron ore in the form of hematite containing about 70% iron. The little lumps look as if one surface has been bubbled, rather like some chocolate, a likeness strengthened by the reddish brown colour of many of the fragments.

However, it was copper and lead that were to become the mainstays of the Lakeland mining industry, but not without conflict. Elizabeth I supported copper mining in her kingdom as it was essential for the production of the armaments needed to maintain her realm's independence from Spain, then the dominant power in Europe. However, there was insufficient local expertise and few incentives to undertake 'innovations'. A report to the Queen noted: 'undoubtedly, there shows so many liklihoods and natural inclinations and shows in these mountains of great store of copper ores and rich leaders. If these were ordered and the several natures of the ores known; how to smelt them with their several additaments, these mines would in time become as famous a mineral town as any in Germany grown up of late years and set many of these idle and loose people to work, whereof this country has great need....we find an unwillingness in the people to discover them. Partly in respect that no recompense is made unto those whose grounds are broken up and, in some parts, spoiled by the stuffs drawn out of the mines; being a people given to make a great complaint of every small loss. Partly for lack of competent rewards to be given to such poor men as would take pains to search out veins'.

Inevitably the need for armaments overcame the local lack of interest. German mining experts were brought to England to share their expertise developed in Europe's most advanced centre of mining technology and techniques. However, the Germans who came to what became known as Godscope mine (near Catbells) were initially received with hostility and suspicion, so much so that they had to live on an island in the middle of Derwentwater. Time and familiarity mellowed attitudes and they were allowed to integrate and marry locally and the local industry developed out of the Germans' expertise.

The practice of mining was not always welcomed. During a dispute in 1620, it was claimed that refining of ore near Coniston had led to flooding and pollution both of fields and streams. The objectors complained that 'both meadow and cornland is decayed and wasted by reason of said stamphouse and brayinge of the copper ore and other rubbish at said stamphouse and then the severinge of the same doth so muddy and corrupt the water wich over-flowing aforesaid ground leaveth such corruption upon them, there is utterly decayed and wasted of the hay and the aftergrass...fishing is utterly destroyed and banished in the said beck or river only by reason of the corruption which cometh down the water from the said stamphouse'.

Another unpopular consequence was that locating the mineral veins in the first place caused damage to the sheep pastures. Some mineral-bearing veins (the earliest ones to be worked) outcropped at the surface, but as they were worked out it was necessary to discover new ones. It was also found that a vein might suddenly disappear (due to faulting) and a search had to be made to find where the veins re-occurred. The technique used to locate veins was to build a dam at a potential site then, once sufficient water had been collected, the water was released to flood the area. The flood of water carried away the surface soil and vegetation, exposing the rock outcrop and, hopefully, a mineral vein.

Goldscope mine in the Newlands valley, first exploited with the help of those German miners, was one of the earliest copper mines. Records show that it was active in the early 13th century. It also produced very small amounts of silver and, much later, lead veins were also exploited around the original site.

Photo 8.6 (top) | Copper mines (left foreground) and slate workings (centre to upper right), Conniston..

Photo 8.7 (bottom left) | A copper mine entrance, Catbells.

Photo 8.8 | Derelict water pipe used to supply waterpower to copper workings – the pipe can be seen running up the slope to the crags at the top.

The mine expanded to new sites in the same area and was joined by other copper mines near Borrowdale, Buttermere, Coniston and Caldbeck to the north of Skiddaw. Lead mines were opened at Loweswater, Threlkeld, Keswick, Helvellyn, Eagle Crag, Hartsop, Myers Head (north of the Skiddaw massif) and Greenside, the latter becoming the largest lead mine in the area. At Coniston, the mines went as far as 500m underground following the narrow veins of minerals. The industry became an important one in an area which depended otherwise in large part on a subsistence economy.

The first sign of an old mine is usually a waste tip, often terminating at a gash or small hole in a rock face (the mine entrance). Usually, after noticing one tip or mine entrance you start to see others in the same area. The waste tips are often worth mooching through to find samples of metallic minerals and a fascinating variety of coloured and oddly shaped quartz fragments. The mine entrances, however, are best avoided. You may go in, but you may well not come back out again. Even walking around mining areas you should take great care where you put your feet as there can be hidden holes and deep water-logged areas into which you could fall. However, if done carefully you can find many remnants of earlier mining operations.

Some of the most spectacular remains are to be found near Greenside mine and the Coppermines area of Coniston. At Greenside, the course of a flue nearly 1.5km long rises up the fellside from the mine buildings to a spot where the fumes were released into the atmosphere. The work done cutting the course of the flue was also intended to try to locate a valuable seam of lead that had been truncated underground by a fault. The flue itself served the purpose of providing sufficient length for any lead caught up in the fumes to precipitate onto the flue for its later recovery (and thus, every so often the flue would be opened up and the precipitate scraped off for reuse). The mine closed for good in 1959. Most of the Lakeland copper and lead mines had closed well before then, the industry reaching its zenith in the 19th century.

The copper working (and slate extraction) at Coniston have combined to create one of Lakeland's most ravaged landscapes with a complex mass of largely dead and decaying artefacts of the copper and slate industries. I say largely because some operations take place today, as is evident from the dusty white tracks that rise up into the mines area.

Another extractive industry was, and to some extent still is, slate quarrying and mining. Although slate can be used for many purposes, its main economic attraction is as a roofing material. Good slate splits very thinly and cleanly along the cleavage plane making material that is light enough to be used for roofing while still being pretty weatherproof. We saw in Chapters 2 and 4 how fine-grained sedimentary rocks were transformed into slate by the

Photo 8.9 | The flank of Fleetwith Pike with Honister Pass on the left. The layers of lavas interspersed with pyroclastic and sedimentary rocks are clearly seen. Slate mine entrances and waste tips are just about visible, although it is hard to distinguish slate waste from natural scree at this scale except by slight differences of colour.

pressures exerted by continental collision. These forces caused all the flattish minerals in the rocks to align their faces in the same direction. However, while many rocks are susceptible to being turned into slate, not all slate splits thinly enough to make suitably lightweight roofing material. In addition, large amounts of rock have to be extracted to get at the good slate, so slate operations are inevitably associated with waste tips (or noisy crushers in some modern sites, such as above Honister).

Slate is known to have been taken from the fellsides since at least 1643 when records first mention slate from Honister. The demand for slate increased substantially in the 19th and early 20th centuries as England's towns and cities grew ever larger. From the early 19th century, slate was transported around England via lakes, rivers and the network of canals (being moved to the water by horse-drawn sledges). Slate quarries or mines opened at Coniston, Tilberthwaite, Troutbeck, Elterwater, Kirkstone and Borrowdale and some are still in operation today. They are often marked by waste tips, quarried mountainsides and the remains of tramways and inclines which were the means of moving the heavy slate to the lakes, rivers and later the railways. The level of demand for slate grew enough to justify investment in these major engineering feats.

Photo 8.10 |

Greenside mine –

disused workings.

There are some good examples of small-scale slate mines and caves to be seen on Loughrigg Terrace at the foot of Loughrigg Fell, although one cave has recently been closed to public access following rock falls (see photo 4.9). Several small-scale workings can also be seen close up on Lingmoor Fell as well as more distant views of large-scale operation (Walk 13). Around Honister in particular the remains of old mining and quarrying operations have also left us the remains of the old means of transporting the heavy slate, such as the self-acting incline above Honister Pass (Walk 2).

The Lake District's towns started to grow significantly from the 17th century and were sites for trade and exchange with markets and fairs. The lower parts of the buildings of Ambleside and Keswick town centre have been so changed by modern shop frontages that it takes an effort to see the underlying buildings. The architecture of Cockermouth however, originally founded in the 13th century, has been indelibly marked by the 18th and early 19th centuries not least with the industrial buildings that line the river banks. Trade, industry and a newly expanding British empire brought some degree of growing prosperity to this remote rural area. One of Cockermouth's most celebrated residents was Christian Fletcher who led the mutiny on the Bounty in very distant oceans, half a world away from the Lake District. This illustrates the widening horizons of residents and increasing integration of a previously remote area into a national and global economy.

Ambleside also grew significantly towards the end of the 18th century, becoming a major tourist destination as more and more people visited the area to see its frightful mountains. Until the second half of the 18th century, visitors to the fells were mainly scientists studying the biology and geology of the area. Tourists soon started arriving in ever greater numbers. Only the brave dared venture beyond the Jaws of Borrowdale or up onto the fells, but their numbers increased steadily and the fellwalking tradition took off.

By the end of the 18th century, the Lake District was becoming one of England's most popular tourist districts and in 1770 the first guidebook appeared; many more were to follow. In 1778, the first of an eventual ten editions of Thomas West's Guide to the Lakes was published. He recommended Castle Crag in Borrowdale as a viewpoint: 'From the summit rock the views are so singularly great and pleasing that they ought never to be omitted ... This truly secreted spot is completely surrounded by the most horrid, romantic mountains that are in this region of wonders.'

William Wordsworth was born in Cockermouth in 1770. His poetry immortalised the area and was part of the intellectual environment which helped open minds to nature as a thing of delight and beauty rather than of foreboding and the unknown. Wordsworth described the area as 'a sort of national property'. He was author of his own guidebook to the Lake District, published in 1820. It is said that it sold far more copies at the time than editions of his poems.

Specialist genres of guidebooks also appeared in the 19th century, including geology and field studies guides. Geological tourism was a small, but not wholly insignificant, sub-culture

Photo 8.11 (top) | Honister slate quarries and mines.

Photo 8.12 (left) | Sedimentary structures in slate of the Seathwaite formation volcaniclastic sandstone.

of the overall tourist trade as more people wanted to learn something about the landscape they sought out for its beauty and tranquillity.

Just as today we refer to a volume of Wainwright's Pictorial Guides simply as a 'Wainwright', in the second half of the 19th century the most popular guidebook was popularly known as 'Murray's', the full title being Murray's Handbook to the English Lakes. It was 'found in the pocket of every self-respecting walker and connoisseur of scenery' who wanted more than just to 'contemplate on the sublime or become rapturous over the picturesque'. Seven editions appeared over a span of 50 years.

Murray's guidebook was intimately connected with the growth of the railways. General tourism grew explosively in the mid-19th century with the coming of the iron rails to the region. Windermere was connected to the train network in 1847, Coniston in 1859 and Keswick in 1864 (today, only Windermere is still 'served' by dirty, often overcrowded and, in my experience, always late-running trains). Key walking routes developed near the stations as well as places for visitors to stay, eat and otherwise spend money.

Towards the end of the century the National Trust was formed and it soon started to acquire land in the area, both open fells and farmland, to ensure public access. The Trust acquired the area around the summit of Scafell Pike in 1925 followed soon after by parts of Great Langdale. One of the most active of the Trust's early stalwarts was the historian, G M Trevelyan. He was once an immensely popular author, selling very large numbers of his books on English Social History and earning very good royalties. With these, he acquired land in the Lake District and gave it to the Trust. He badgered the good and the great to do the same and we owe not a little of the open fells and present day NT-owned low land to his untiring efforts. He was also an indefatigable walker, often taking unsuspecting guests for a 30 mile stroll after lunch.

Trevelyan and others also campaigned against development in the Lake District. In 1912 he opposed plans for a road to cross over the Scafell peaks and in the 1920s lobbied against a proposal to build a road into upper Eskdale. Trevelyan was also active in the Youth Hostel Association and the Ramblers Association. The YHA set up a network of 'spartan' hostels, each about a day's good walking apart, and the Ramblers set about working for improved access. Trevelyan also encouraged another great walker, Hugh Dalton, a minister in the 1945–1951 government, to introduce National Parks into Britain. It is to the conservation and access work of Trevelyan and contemporaries that we owe the largely undeveloped fells of Lakeland with their open access. However, it would be a mistake to view Trevelyan as a sort of proto-modern environmentalist. His concerns were primarily with the loss of aristocratic

land-owning and social structures under the pressures of economic and social change generated by industrialisation. However, we owe to him and his ilk a great debt for preserving the
area from unthinking development in the early 20th century.

Today Ambleside, Windermere, Grasmere, Keswick and Penrith are thriving tourist centres,
their streets thick with de-coached coach parties, car-borne day trippers and fellwalkers seeking a cup of refreshing tea, shelter from the rain or a new pair of boots from the innumerable
outdoor gear shops.

The post-war period also saw another dramatic development: the growth of car-based
tourism. Soon the roads that were once the arteries for walkers and cyclists had become the
exclusive preserve of the car. The traffic speeds and density had made it so unpleasant and

Photo 8.13
One that got away –
an undeveloped old
road near Ambleside.

Photo 8.14
Upper Eskdale and
the central fells.
Left to right: Scafell,
Scafell Pike, Ill Crag,
Great End, Esk Pike
and Bowfell, seen
from Border End.

dangerous to walk on the roads themselves, that in the 1970s, the National Trust found it necessary to open up footpaths on some of its properties to run behind the old stonewalls alongside roads (Walk 11 in this book makes use of such a footpath on the eastern side of Derwentwater). Some old lanes still survive in the form they had 100 years ago and more, but they are few and far between and today the roads are dense with motor traffic all year round.

I have a copy of a book published in 1970 about the Lake District which includes a photograph of the main road from Ambleside to Keswick jammed solid with traffic. What is striking to the modern eye is the small size of the cars compared with the ever greater size of modern cars (even although we live in supposedly more environmentally aware times). The jams we take for normal.

The authors of the book said that while they regretted the number of vehicles it was still possible for the walker to get away from it all. They wrote that once you left the traffic-drowned main road from Ambleside to Keswick and got up onto Nab Scar on the way to Fairfield (Walk 10 in this book), you were very unlikely to see another person. Today it is the other way around; on this great walking route (the Fairfield Horseshoe) you are most unlikely not to see plenty of other people.

The much-loved author of fellwalkers' guides to the Lake District, Alfred Wainwright, adored the fells for the solitude he found there and for the fulfilment he gained from working on his marvellous drawings and wry commentaries. Today his guides are a major industry in themselves, with the publishers thinking up ever newer wheezes to cash in on his name. The original guides easily remain the best of this massive output and still help direct many thousands across the hills.

Yet solitude is now just about the last thing you will find, certainly in the central fells area (except perhaps in remote upper Eskdale). No doubt the number of visitors would be no less even if Wainwright had not written his books. But if it is solitude you are seeking, then the central fells are not the place to search for it. People come and in ever greater numbers. Yet, those millions still love the fells with the same fierce intensity as did Wainwright. People now tramp the fells for the pleasure they gain from walking and climbing, from the effort and from the views of the landscape that reward their effort.

My hope is that learning something about the geology of the fells, and therefore appreciating the scenery more intensely, might mean that knowledge gained is a substitute for solitude lost.

About the Walks

Two factors have been central to selecting the walks: quality vistas and geological/geomorphological interest. I have also selected the walks to include a range of rock types, although there is an emphasis on some of the main pyroclastic rocks which dominate the central fells area. I've tried to cover the most popular summits as well as some less well-known ones. However, I've also had to leave out some of my favourites, such as Coniston Old Man and Blencathra.

The walks require differing levels of stamina, ranging from fairly easy to some quite long and demanding walks. Where practicable I've indicated when a long walk can be cut short. I should emphasise that some of these walks are tough, potentially dangerous routes in remote, crag-ridden areas, and require good navigational skills if the weather changes. So, unless you are confident that you have the skill to navigate in misty/cloudy conditions (perhaps in high winds, rain and hail), the higher level walks should only be tackled in fine conditions.

There are low level walks for days, frequent in Lakeland, when, even if it's not raining cloud obscures the summits. However, cloud and mist can usefully focus attention on the 'micro-landscape'. The low level walks offer some options for cloudy or wet days and, unless the cloud comes down uncommonly low, still give good views. These lower level walks range from fairly straightforward to hard-going. They are designed to explore a different scale of mountain landscape. Walk 15 is slightly different and is included as it provides a fairly short walk that allows the walker to see a wide range of pyroclastic rocks within a small area.

The sketch maps for the walks are designed to show the general route and are for use in conjunction with the OS 1:25,000 maps (OL4, OL5, OL6, OL7). My maps are very emphatically not designed to be sufficient to use on their own as it is impossible to show the precise route with sufficient accuracy. It is therefore essential, before going out on the mountains, to plan the route on the OS map and to take it with you. Grid references are given by six or eight figures as appropriate; for the walks most, but not all, have been verified using GPS.

The OS 1:25,000 maps are incredibly good value, each covering a considerable area. The only complaint is that you need four of them to cover the whole of Lakeland and many walks in the central areas require two or more maps where they cross from one to the other.

The OS 1:63,360 (one inch to the mile re-issued by the OS as 'Touring Map 3' in 1998) and the Harvey 1:40,000 Lake District map are useful for showing the shape of the ground better than the 1:25,000 map, where the amount of detail can obscure the contour lines that allow the shape to be interpreted from the map. While that makes them helpful as an adjunct, the detail of the OS 1:25,000 map can be vital in restricted visibility, so if you only take one map, make sure it is the appropriate OS one (and preferably a waterproof version).

It will make sense to read the description before going on the walk and then to take a photocopy, or the book itself, on the walk as a reminder of what to see and where.

To give the reader an idea of how hard a walk might be and what sort of challenges it may pose, I have borrowed (and slightly adapted) a system for ranking the walks from Ralph Storer's book, *100 Best Routes on Scottish Mountains*. This system allows a more nuanced assessment to be given than 'hard' or 'easy'. For each walk I give a ranking on a scale of 1 (easiest) to 5 (hardest) for: navigation, terrain and severity. Note that a walk may be very easy navigation or terrain-wise, but still have one or two short but seriously more challenging sections. In such situations I have given the ranking earned by the most difficult section.

A high navigation ranking means that there are places or sections on the route where it would be easy to make an error and head off in the wrong direction towards potentially dangerous ground.

A high score for terrain may be due to factors such as having sections of steep ground, the need for some scrambling (i.e. the essential use of hands as well as feet to ascend/descend) or it may be due to some very rough going (usually tussocky grass/heather and hidden boulders and holes in the ground).

By 'severity' I mean the chance for things to go badly wrong in the event of a minor navigation error or a simple accident. In general, the most serious walks are those that go high and/or off the popular tracks to remote areas with few other people around.

I have used some terms in specific ways. In general I have used the term 'track' for any obvious trail on the ground from a motor-vehicle track to a sheep track, and anything in between. Some tracks are marked on the 1:25,000 OS maps by thin dashed black line (which can be hard to see), but many more are not marked at all. Terms such as narrow track, wispy track and twisting track are all, I hope, self-explanatory.

By 'footpath' I mean a public right of way marked in green on the OS 1:25,000 maps (thus including bridleways). It is important to note that the course of the actual track on the ground may not be the same as the line of the footpath shown in green on the map; this can be confusing if navigating in an unfamiliar area and should be watched out for. Only a very small proportion of tracks on the ground are also footpaths, so most tracks are not marked on maps.

In the directions I use phrases such as 'after about 10m of descent' (or ascent) to indicate vertical distance to be travelled and phrases such as 'after another 100m or so' to mean horizontal distance. The former will usually apply on steep ground and the latter in less steep areas, although it depends on the specific circumstances – for example, with a gently sloping track diagonally crossing a steep slope, distance would probably be appropriate.

A 'tarn' is a mountain lake (but not a big valley floor lake which is actually represented by the 'mere' in some lake names); a 'gill' is a mountain valley; a 'dale' a slightly bigger valley; a 'beck' is a mountain stream; a 'hause' (pronounced 'hows') is a pass or col.

As for place names, I have generally followed those used on OS maps with one major exception: I use the more common Scafell rather than Sca Fell preferred by the OS. Two points about place names are worth mentioning. First, the same name is often used two or more times for different places in Lakeland, for example there are several Angle Tarns and Sour Milk Gills. Sometimes the same name is used for different places but in slightly different spellings, for example, Whiteside (north-western fells) and White Side (Helvellyn range). Second, one place may have more than one name – Hopegill Head is also known as Hobcarton Pike, and High Pike is sometimes known as Scandale Fell. Here I have used the name given first on the OS maps. Sometimes names are used in different ways. Buttermere, for example, is a lake, but at times is also used to mean the valley and/or the small cluster of pubs, hotels and farms that form the hamlet of Buttermere. Usually the precise meaning is clear from the context.

Also note that the different scale OS maps and the Harvey 1:40,000 map often give different names. For example, the OS 1:25,000 and 1:63,360 maps name the hilly area to the south-east of Ullswater's most northerly section as Barton Fell and has the names of several individual high points (including Arthur's Pike and Bonscale Pike) in equally small type. However, the Harvey map omits the name Barton Fell as well as several of the less high individual points and puts Arthur's Pike and Bonscale Pike in heavy type. This sort of thing occurs often enough to be a nuisance to the guidebook writer, and no doubt to the hillwalker too!

Getting to the walks

I have tried to ensure that all the walks are accessible by bus as well as having nearby car parking facilities (although parking anywhere in the Lake District is quite expensive). Unfortunately, Cumbria County Council's website on public transport is pretty much useless unless you already know the bus number of the service you need to look up. (It is useful, however, if you want to read loads of pointless self-serving puff about what the council is doing to promote bus use.) The best bet is to download the bus service map (a link can be found at www.cumbria. gov.uk/roads-transport/publictransport/busserv/ but beware as it is 5MB in size), apply for one by post or, if already in Lakeland, try to acquire one from a Tourist Information Centre. The map shows bus routes and numbers. You can then use the web to check bus times.

Hazards

It is possible, even for the experienced, to underestimate the physical demands of walking on the Cumbrian mountains and of the potential for good weather to turn into threatening conditions, sometimes within minutes. Mid-summer can feel like winter if you are high up and the cloud, wind and rain descend. Always take spare clothing, water- and wind-proofs, a torch, food and drink, a good quality survival bag (such as those available from Blizzards Protection Systems) and a map and compass. On your feet, you need good boots. Check the weather forecast before you start (on the Met Office website mountain area forecasts page).

It may well be possible to go up and down the fells without a map, relying instead on a Wainwright guidebook or even just a leaflet from the local tourist office (thousands do so every year). However, things can easily go wrong if you are not fully orientated and aware of the scale of the mountain landscape.

One day while researching this book, after dropping down off Pike o'Stickle the mist descended while I began the short climb to the summit of Harrison Stickle. Having seen that conditions were worsening I had made sure I knew where I was on my map. Later on, after a cup of tea and a look at the rocks on the summit, I was about to descend to Stickle Tarn when I was met by two walkers. They were using a Wainwright guide but not a map and had become confused about the number of cairns that the guidebook indicated should be on the summit area. As we started to descend we were approached by a couple more walkers (wearing shorts and trainers) who were also mapless and wandering about in confusion.

It's very easy on complex mountain terrain, even in clear conditions, to mistake your position and head off in the wrong direction unless you have with you a map and a compass and, crucially, knowledge of how to use them. In misty conditions, an insignificant gully can easily look like a major gorge. Unless you are aware of your position on the map, it is extremely easy to wander unawares onto dangerous ground or into the wrong valley and end up miles from where you want to be.

Remember that if you go high up you must be prepared to navigate your way off the mountain using map and compass alone if, for example, mist should descend suddenly while you are on the summit.

Modern technology has great potential for hillwalkers: the mobile phone and GPS. Mobile phones have meant that mountain rescue teams now get called out about two hours earlier than used to be the case. This means casualties get out of adverse conditions or are taken to hospital earlier and mountain rescue members (all unpaid volunteers) get fewer late-night call-outs. On the other hand, the mobile has also led to an increase in the overall number of call-outs, sometimes for silly reasons. To save the volunteers from unnecessary call-outs, only use your phone in a real emergency when you should dial 999, ask for the police and then tell the police telephone operator that you need mountain rescue (rather than asking for an ambulance). Don't rely entirely on your mobile as you may well not get a signal, so do take a whistle as a back up (six long blasts every minute is the recognised distress signal).

Similarly, handheld GPS can be mightily useful, but needs to be used sensibly and in conjunction with map and compass, not as a substitute for them. In 2007, Lake District mountain rescue association reported two call-outs where the callers were lost despite having a GPS system with them, because they hadn't bothered to learn how to use it.

No skills are of more use to hillwalkers than navigation and map-reading. There are many instructors and organisations within Lakeland and elsewhere who provide training in navigation, first aid and other mountain skills and who can be discovered after a few minutes web-browsing. Unfortunately, there doesn't seem to be a single website to use as there is in Snowdonia, so it is necessary to search around a bit to find a choice of course providers.

For the landscape lover, the usual mountain hazards are multiplied. Wandering off route to take in a view can add to the time needed for a walk and this should be taken into account, especially in the short days of winter. The times suggested for the walks in this book are therefore somewhat lengthened to allow for stopping to explore rocks and features, contemplation and refreshment, enjoyment and, hopefully, enlightenment.

Exploring old mines and quarries is particularly dangerous. Do not enter mines, tunnels or holes in the ground. In copper and lead mining areas especially, look out for open holes in the ground down which you could easily fall, never to come back up again. In quarries watch out for unstable rock faces which may collapse at any time. The same is true for waste tips which can be unstable, especially after heavy rain. Don't get too close to the edge at the top of quarry faces – those little gullies running parallel to the cliff edge are tension cracks where the rocks are preparing to slip. From below, always look up as you approach a cliff or quarry face to see if there is anything that looks as if it is about to topple on you.

Geology maps and general guides

A geology map is extremely useful if you want to become more familiar with the different rock types. The British Geological Survey publishes two 1:50,000 maps covering most of the Lakeland area: the Keswick area (sheet 29) and the Ambleside area (sheet 39). The Appleby area map (sheet 30) includes the High Street area, but mainly covers the area to the east. If you buy such maps make sure you specify 'Solid' (this term is to be replaced on future editions by 'Bedrock'). Other available types, while offering more information, are too detailed to be practical in use and those with the terms 'Solid and drift' or 'Solid with drift' (to become 'Superficial' in future editions) are best avoided.

Bryan Lynas's *Rocky Rambles in Lakeland* (Sigma, 1994) is well worth reading if you can find a copy (it is now out of print). I have devised the recommended walks in this book as much as possible to complement, rather than to repeat, the walks outlined in Lynas's book but there has been some inevitable minor overlap.

For general guides to the fells, the famed Wainwright books remain the gold standard but buying all six books would cost a small fortune. W A Poucher's *The Lakeland Peaks* deals with all the main ranges with excellent photos, many with the routes marked on them to help with planning and navigation.

As matters on the ground can change (stiles may be moved or new fences appear) some of the information about the walks may become out of date. I will try to post updated details if or when I become aware of them on the *Rock Trails Lakeland* page of the website *www. rock-trails.co.uk*, where readers can contact me with their own updates/information.

Entreaty

Please take note of the Countryside Code. The most important injunctions are to close gates and leave absolutely no litter including so-called bio-degradable matter such as tissues, tea-bags and orange peel. For geologists and landscape walkers there are a few more things to think about, listed in the Geologists' Code. As this is a National Park one should not even remove loose items from quarry/mine waste tips and certainly there is no reason at all to try and lever or smash samples out of rock outcrops. For everyone, the main thing is to use common sense in minimising the impact you have on the environment.

Walk #1

Gasgale Gill & Hopegill Head

START	▶	LANTHWAITE GREEN 159 207 (BUSES SERVE BUTTERMERE)
FINISH	●	CIRCULAR ROUTE
TIME	⏱	5 HOURS+
GRADE	⊙	NAVIGATION ● ● ●
	☁	TERRAIN ● ● ●
	✪	SEVERITY ● ● ● ●

This walk takes you through and over some of the most dramatic of the scenery created by the sedimentary rocks that dominate the northern and north-western fells, collectively known as the Skiddaw Group. Skiddaw Group rocks make up popular peaks such as Grasmoor and Causey Pike, as well as Skiddaw and Blencathra on the Skiddaw massif.

Gasgale Gill & Hopegill Head

I have picked this walk to introduce the Skiddaw Group for several reasons. The Skiddaw Group rocks in themselves are not overly exciting, lacking features that make for easy identification (indeed geologists point out that the Skiddaw Group sedimentary rocks look very much like some of the sedimentary rocks found within the Upper Borrowdale Volcanic Group). However, on this walk you encounter a wide range of geological features, including the line of a major 'thrust' fault, bedding, slumped bedding, slaty cleavage and folded rocks. These features come with excellent views and good walking.

The walk follows two outstanding scenic gems: the intensely enclosed atmosphere of Gasgale Gill, especially in its lower reaches, and the fantastically airy ridge walk between Hopegill Head and the western end of Whiteside. Both are expressions of important geological structures and processes.

Gasgale Gill partly follows the line of a fault, while on the valley slopes to the north of the beck there is a major thrust fault. Along the ridge top there is evidence of convoluted bedding due to intense folding and thrusting of the rocks. Right on top of the ridge, just below the summit of Hopegill Head, you can see the axis of a syncline in the rocks. The ridge line is also a boundary between two different 'formations' within the Skiddaw Group. There is also indirect evidence of a massive underground intrusion of granite.

The Skiddaw Group used to be known as the Skiddaw Slates. While the Skiddaw Group is made up of a variety of sedimentary rocks, many of which display slaty cleavage, not all of them do so. Even those Skiddaw Group rocks which have been affected by cleavage are not worked commercially. The slate extracted from Lakeland quarries is taken from sedimentary rocks interleaved between the volcanic rocks of the Borrowdale Volcanic Group. So, although the rocks seen on this walk are 'slaty', they are not exploited for slate.

The walk starts by going up Gasgale Gill (a deep etching into the mountains on either side), Hopegill Head and Whiteside to the north and Grasmoor to the south. (Those starting from Buttermere could instead miss out Gasgale Gill and reach Hopegill Head via an ascent of Grasmoor.) The Gill is easily viewed from the surrounding high land, especially between Hopegill Head and Whiteside. Although many of the macro-features can be seen from above, a walk through the Gill is an entrancing alternative approach showing an all too often neglected aspect of the geology and scenery of the area.

Gasgale Gill & Hopegill Head

From the start point, head roughly west up an obvious track towards the start of Gasgale Gill, aiming for the footbridge at 163 209. Cross the footbridge and turn right to follow the lower level path into the narrow entrance to the gill with the beck on the right. Initially the gill is very steep-sided, pressing in almost canyon-like, providing a highly atmospheric start to the walk. The narrow valley is almost claustrophobic in intensity especially when the beck is in spate and it is difficult to avoid having one's attention drawn back and forth between the speeding waters and the looming crags overhead (see photo w1.1). Some minor scrambling is necessary at the beginning in one or two places and care is needed, but any slight difficulties are soon over.

Photo w1.1 | Gasgale Gill, with smooth slabs exposed on one side of the valley (looking downstream).

Photo w1.2 | Gasgale Gill, with rough, craggy edges exposed on the other side (looking upstream). Hopegill Head is in the centre distance.

Gasgale Gill & Hopegill Head

The crags on either side of the gill are tilted in the same general direction, from lower left to upper right as you turn round and look back down the valley. This can be seen in the different shape of the outcrops on either side in this narrow section. To the right (south) the beds are exposed, tilted at about the same angle as the slope and leaving a smooth surface. However on the left, a cutting across the beds is exposed, leaving a craggy surface as each has been broken roughly through. This is illustrated in photos w1.1 and w1.2; note, however, that photo w1.1 is looking downstream while photo w1.2 is looking upstream.

In this narrow section, keep an eye on the outcrops to the left. Bedding can clearly be seen in some of the rocks forming the lower parts of the crags above. In some places the bedding has clearly slumped, the result of earthquakes causing downwards sliding of still not fully solidified beds of sedimentary deposits.

After a while the gill opens out, losing its deep canyon-like aspect due to a shallowing of the valley sides, especially on the southern side below the glacial corrie and cliffs of Dove Crag below Grasmoor. This is caused by the changing underlying geology with softer rocks being exposed in the widened area of the valley.

Higher up the gill, the side walls again begin to close into a sharp V-shape and there is an obvious knick point at a rock step across the beck, producing a waterfall. Here is a good point to turn around, if you haven't already done so, to look down the valley towards Gasgale Crags on the right. The base of the crags marks the line of a major thrust fault, the Gasgale Thrust, that we will mention again later on when talking about several linear features running through the gill.

Shortly after resuming the upward slog you arrive at Coledale Hause, a complex col linking Crag Hill and Grasmoor to the south and Hopegill Head and Grisedale Pike to the north. In good weather there is no real navigational challenge in finding the track up Sand Hill, a prominence just below Hopegill Head, and on to the summit.

The summit of Hopegill Head (sometimes known a Hobcarton Pike) is a fantastic viewpoint, with ridges heading directly off to the west (Whiteside), north (Ladyside Pike, see photo 1.7) and east (then bearing north-east) to Grisedale Pike. Other mountain ridges roll off in the distance.

If the weather permits, the summit is an excellent viewpoint at which to stop and study the sedimentary landscape around you. However, in windy conditions this is an exposed spot and shelter can be hard to find.

Studying the surrounding hills, it is possible to see that the general trend of the landscape is roughly south-west to north-east, with ridge lines running from the Buttermere-

Crummock Water valley to that of the River Derwent. There are some secondary arms running north/north-west, but in general the trend of the hills lies across the direction from which the mountain-building pressures came 300 million years ago. The underlying tilted and folded beds have been cut across to produce the present fells, like solid waves. In general these waves, expressed as the long, narrow-ish ridge lines which provide such super walking in the north-western fells of Lakeland, reflect the tilt of the beds: the north-western slopes run up the bedding and the south-eastern slopes cut down through the beds.

Looking west along the ridge towards Whiteside (photo w1.3), two different formations within the Skiddaw Slates make up the northern and southern slopes. The ridge itself and the southern slope, down to Gasgale Gill, is part of what is known as the Kirkstile Formation, consisting of beds of mudstone, siltstone and sandstone. The rocks on the slope to the right of the ridge are known as the Loweswater Formation of sandstones originally laid down in water and subject to slumping caused by earthquakes. The Loweswater Formation rocks come again to the surface near the Gasgale Beck and form the slope running up to Grasmoor for a short distance.

Photo w1.3 Looking west along the summit ridge from Hopegill Head towards Whiteside.

About 10m west of the summit, on the very ridge line, some evidence of folding of the rocks is exposed. There are clear signs of bedding and slaty cleavage. If you get down close to the rocks and look back up them to the summit, you can see that there is clearly a syncline (or downward fold) in some of the rocks (see photo 4.2). In fact, a complex set of anticlines and synclines runs through this area from east to west, some coming up through Gasgale Gill and onto the summit of Hopegill Head and the crags below Sand Hill.

The gill was cut in pre-existing drainage channels by powerful glaciers in (geologically) recent times compared with the ancient mountain-building episodes. However, it is possible that some of the slopes that can be seen from here (e.g. the summit area of Grasmoor and Causey Pike), are probably remnants of the pre-glacial surface that have survived above the valleys we see today.

Looking back down into Gasgale Gill, and starting with the slopes below Grasmoor, the glacial corrie below Dove Crag is quite clear (photo w1.4). The corrie is not very deeply etched, nor does it contain a lake. When temperatures began to rise towards the end of the ice age, this north-westerly located corrie was one of the first to see its glacier melt away. It thus remains fairly undeveloped compared with, say, Blea Water on High Street where the ice melted much later.

Less obvious is a series of linear features including an anticline which runs through the northern end of the Dove Crag, a set of very closely spaced anticlines and synclines which run up to and through Sand Hill and Hopegill Head and others which cut through Gasgale Crags.

One major feature is indirectly visible: the line of the Gasgale Thrust we saw earlier on exposed Gasgale Crags below Whiteside on the northern side of the valley. This continues through just north of Hopegill Head and the northern slopes of Grisedale Pike.

A great wedge of lower (and thus older) beds of the Kirkstile Formation (slumped and folded siltstone and sandstone) has been thrust upwards here to rest against younger beds of the same formation (siltstone and mudstone) as a result of mountain-building forces when the beds of sedimentary rocks were folded. The cohesion of the rocks was destroyed, creating both the complex set of anticlines and synclines in this area as well as the Gasgale Thrust fault. (There is another major thrust fault, the Causey Pike Thrust, which runs just about 1.5km to the south.)

The thrusting here has caused older sedimentary rocks, which form the northern slopes of Hopegill Head and Whiteside, to have been shoved up above the newer rocks which form the southern slopes running down to the beck.

Another geological feature is also indirectly and indistinctly visible, a 'metamorphic aureole'. This marks the limit of the area of rocks heated and toughened by the intrusion, well below ground level, of a mass of molten rhyolitic magma which slowly cooled to form granite. Heat radiated up through rocks in contact with the hot intruded magma, toughening them.

Photo w1.4 | Grasmoor and Gasgale Gill: the immature glacial corrie can be seen in the upper centre left of the photo above the green slopes.

The limit of this effect is marked by a line on the southern side of the valley, running from just south of Coledale Hause roughly along the base of the crags opposite where you are, via the northern edge of Dove Crag and along to the narrow entrance to Gasgale Gill which you earlier walked through. Indeed, this metamorphic aureole largely explains why the gill is gorge-like at the beginning and then widens out. The toughened rocks to the south of the line have been carved into crags below Grasmoor which is nearly 100m higher than the surrounding fells, whereas the softer rock north of it, running up to the beck, has been more easily eroded into shallower slopes.

All of this is quite a lot to take in from one spot, but it's worth getting it all sorted out before carrying on along Whiteside. There are some other features which will attract the attention, not least the marvellous views along this great ridge walk which lasts for 1.6km.

Try to observe the different rock types as you walk along the ridge, noting especially the cleavage which is clearly seen in places. There seems to be no pattern to the different directions of the cleavage in the various outcrops which lie at all sorts of angles, even vertical. This is the result of the complex patterns of folding and thrusting which has distorted the original patterns.

Gasgale Gill & Hopegill Head

Photo w1.5 | View back along the ridge from Whiteside to Hopegill Head (top right).

Further along the ridge, especially in Gasgale Crags, you can see more examples of the slumping and folding of bedding as well as more cleavage. In some places you can see examples of both bedding and cleavage in the same rock exposure, but these two features can sometimes be difficult to distinguish (see photos 1.4 and 4.3). Again it is all but impossible to work out any pattern to the tilt of the bedding due to a complex of anticlines and synclines cutting through the crags, roughly running from east to north-west.

Eventually this delightful ridge walk must come to an end and, after reaching the western summit of Whiteside, there begins a steep descent through numerous interesting outcrops where further examples of bedding and cleavage can be seen. At about 1690 2158 (well to the east of the footpath marked on the OS map) the track crosses the Gasgale Thrust fault line and, by looking back up the gill from the edge of the crags, you can see the line running along the base of Gasgale Crags. The terrain also becomes a bit smoother and noticeably less craggy from here on.

Continue to descend until you reach the footbridge crossed at the start of the walk and, after crossing it, walk the short distance from there to the finish point at Lanthwaite Green. Alternatively, if returning to Buttermere, do consider taking in Rannerdale Knots if you have the energy.

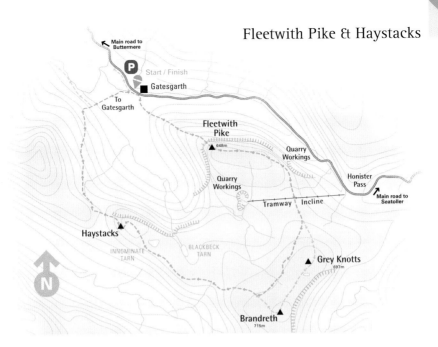

Walk #2 Fleetwith Pike & Haystacks

START	▶	GATESGARTH, CAR PARKING AND BUS STOP, 196 149
FINISH	⬤	CIRCULAR ROUTE
TIME	◔	5 HOURS+
GRADE	⊙	NAVIGATION ● ● ● ○
	◔	TERRAIN ● ● ● ○
	⊙	SEVERITY ● ● ○

Haystacks is nowadays assured of fame as Wainwright's favourite Lakeland fell. This may seem strange, given the range of loftier and more conspicuous peaks in the area to choose from. The summit area of Haystacks is not especially prominent. Being a relatively low, tilted and undulating plateau, its southern side merges inconspicuously into a broad slope running up to Grey Knotts and Brandreth. But, with Haystacks, it's a case of familiarity breeds appreciation.

Fleetwith Pike & Haystacks

Haystacks is a compelling viewpoint precisely because it is encircled by higher fells. On the southern approach, it is difficult to say for sure when you are on the area between Haystacks and Grey Knotts/Brandreth and when you have reached Haystacks itself. However, from the east, north and west it is a different story with sharp and impressively steep crags. The knobbly summit plateau is also an interesting area, revealing new vistas and hidden dips at every turn or rise. From Honister Pass, the ascent of Haystacks provides a wealth of rewards for comparatively little effort.

For those with an interest in geology and rocks, Haystacks and the surrounding area reveal a wealth of fascinating rocks and other features.

This walk takes you across the boundary between the sedimentary and volcanic rocks (mainly, but not exclusively, lavas) of the Lower Borrowdale Volcanic Group, around the head of one of Lakeland's most spectacular glacial valleys and to the top of two fine but not especially high fells (Fleetwith Pike and Haystacks). Both have rather flattish summits, but the interesting detail and surrounding higher fells make this a fantastic walk.

There is a stiff ascent onto Fleetwith Pike and later on a steep descent from the summit of Haystacks to the col of Scarth Gap, but otherwise the going is fairly straightforward. However, there is probably little point in doing this walk in the mist as route-finding for the described walk could be quite tricky between Fleetwith Pike and Haystacks. If the mist descends while you are on Fleetwith Pike, rather than following the described route, head for the main Honister-Haystacks track (using map and compass bearing) and then follow the track past Dubs Quarry, on to the summit and down to Scarth Gap.

The start point is at Gatescarth where there is a bus stop and parking facilities. If starting in the hamlet of Buttermere you have the advantage of getting to the starting point by following the 'permissive' path along the north-eastern shore of Buttermere.

Take the farthest south-east of the two footpaths leaving from just beyond Gatescarth Cottage. The path climbs steeply uphill, initially to the left of the crest, but after going round the lowest cliff face it sticks firmly to the blunt crest of the ridge.

If you haven't done so already, stop at the cairn you meet a short way up for a view back down over Buttermere and Crummock Water. You can see Haystacks and the deep layers of

lava of which it is formed to the south-west (right as you look up the ridge).

The boundary between the Skiddaw Slates and the lavas of the Birker Fell Formation (the first of the volcanic eruptions in Lakeland) is met at about 2025 1435 where the rock type clearly changes from cleaved to uncleaved rock. As you go higher you cross bands of lavas, each representing a different eruption. You can see flow patterns in the lava and areas of 'brecciation' where the lava has solidified slightly before being disturbed again, thus breaking up in part.

Remember to turn around again for fine views over the valley below (see Photo 6.9). The ridge is an example of an arête, a narrowish knife-edge carved by valley glaciers in both the Gatesgarthdale/Honister valley to the north and Warnscale Bottom to the south. Their combined power, aided by glaciers from the corries below High Stile, carved out the main Buttermere valley. The meltwaters eventually formed one large lake at the end of the ice age. Since then, erosion of material carried down from the surrounding fells by Sour Milk Gill, Near and Far Ruddy Gills and Mill Beck has cut the lake in two as the material was deposited in the lake.

There are also fine views towards the glacial corries below High Crag and High Stile which would have been most active late in the ice age when glaciers only thrived at higher levels. You can also see the different topography carved by the ice: smooth in the softer sedimentary rocks on Dale Head and Robinson on the right, and much craggier on the High Stile ridge.

The summit is soon reached, with some short sections of scrambling necessary. The scrambling is fairly easy except when the wind is very strong, when this section could potentially be dangerous. The summit ridge spreads out rather unexpectedly when you get to the top of the ridge, but is hardly flat, consisting of innumerable minor crags and dips.

The lavas met on the climb up the ridge are replaced by interleaved bands of andesite intrusions, rhyolitic rocks (the result of pyroclastic eruptions) and also some sedimentary beds. You can see very rough cleavage (Photo w2.1) in some of the rhyolitic rocks and also patches of brecciated rock.

These rocks are part of the Eagle Crags Member, a part of the Lower BVG. The Eagle Crags Member and some other pyroclastic rocks were produced intermittently in small eruptions between major lava eruptions, which otherwise dominated the Lower BVG volcanic sub-cycle. These pyroclastic rocks were therefore laid down well before the great pyroclastic eruptions that produced the Scafell caldera and other areas, and which produced the bulk of the pyroclastic rocks in Lakeland.

The Eagle Crags Member is a varied group of rocks, including tuffs, coarse tuffs, breccias and andesite 'sills' (intrusions) as well as siltstone, sandstone and conglomerates. Such a variety of rocks has a variety of origins. There were pyroclastic air-falls and surges as well as 'reworking' (i.e. erosion and sedimentation in lakes or seas of the material from these eruptions). The sedimentary tuffs and reworked material are what produced the silstones, sandstones and conglomerates, some of which were later transformed into slate which has been quarried since the late 17th century right through until today.

If you follow the northern/north-eastern edge of the summit area, you will have superb views over to the Robinson/Hindscarth/Dale Head ridge. Farther along the edge, you may see the top of the new 'via ferrata' (iron way) constructed by the owners of Honister Slate Quarry which, for a tidy sum, now provides the adventurous with the experience of this Alpine-style route on Lakeland's own crags. Along this edge, you can't miss seeing the various quarries scattered over the area between Fleetwith Pike and Haystacks (Photos 8.10, 8.11 and 8.12). The workings seen on the other side of the valley and below you on this side all hint at the extent to which these fellsides would have been intensely busy places when the quarry was in operation.

Around about 215 137, start to move away from the edge. Head roughly south, surrounded by quarry workings, to intercept the main vehicle track (and walking route to Haystacks if you encounter misty weather). You may pass close to the now re-opened Hopper Quarry to admire the contribution to the landscape made by the slate extraction operations

Photo w2.2
View from below
the summit of
Brandreth.

and attendant machinery, of various generations and states of decay. The noisy slate-crushing machines do at least have the advantage of reducing some of the need for waste tips. Carry on heading roughly south to intercept the old quarry tramway near the old drum house (marked on the OS map) at about 215 134. The old tramway was used to pull slate up from Dubs Quarry and lower it down to Honister in rope-drawn trucks on an inclined plane using gravity to balance the loads. Bits of old metal from the drum house can be found around the centre of the causeway remains.

From here, follow the public footpath roughly south towards Grey Knotts and Brandreth. Unless you are full of energy you may wish to avoid the summits, and stick with the footpath to about 213 124 when you strike out north-west (see Photos 2.11 and w2.2). Aim for the corner in the wire fence, and initially towards Great Round How. When the fence allows, head to the south-eastern end of the tilting summit plateau of Haystacks, south and west of Blackbeck Tarn, before heading towards the summit. The object of this southerly loop is to avoid the descent and re-ascent required for a more direct route, but also allows you more scope to explore the summit area of Haystacks by approaching from the south.

In good weather you should be able to see quite clearly the main track to the summit of Haystacks over to the northern end of Blackbeck Tarn and, if you wish, you can make your way to that track and follow it to the summit. It's better to pick your own path through the ups and downs of the summit plateau, however. This allows you to study the rocks which are all andesitic lavas, many of them displaying flow-banding (see Photos w2.3 and w2.4). The flow-banding becomes better defined as you get closer to the summit, and some of the

Fleetwith Pike & Haystacks

Photo w2.3 & w2.4 | Flow banding in lava outcrops near the summit of Haystacks..

best examples are seen just around that area. If you have the time, you can pick your own convoluted route looking at the multiplicity of outcrops on your way to the summit.

Make your way to the summit past the summit tarn nestled in a dip in the rocks. I once camped here, on one of those frustrating days when the sky had been clear all day but the mist descended just as we reached the summit. There was no opportunity to watch the glorious sunset one always hopes for on backpacking trips. In fact, the weather turned steadily worse: the wind picked up and later the rain came in. Our tents were protected from the fierce wind and the worst of the rain by the knobbly rock outcrops. The rain had stopped by the next morning, but it was still extremely windy and cloudy. The plan to continue over High Stile and Red Pike was abandoned for a direct route to the teashop at Buttermere.

The descent to Scarth Gap was extremely difficult because of gusting wind and heavy packs (sometimes it would take a while before you could safely move on the steep steps downwards). However, all this bad weather had an advantage: it focused attention on the rocks. It is well worth paying close attention to the rocks on the summit and during your initial descent to Scarth Gap for examples of highly impressive flow-banding in the rocks. I have been back to Haystacks a few times since and have been lucky enough to enjoy better views of the neighbouring fells – but beware: despite being a small mountain, Haystacks can be quite dangerous in very bad weather.

From the summit of Hystacks, look at Fleetwith Pike in clear conditions, where you can see the layers of lava tilted from lower right to upper left topping the sedimentary rocks. Then start the steep section of the descent. From Scarth Gap follow the track either back to Gatescarth (by crossing the valley floor) or to Buttermere (by keeping on the west side of the lake to the northern end of the lake).

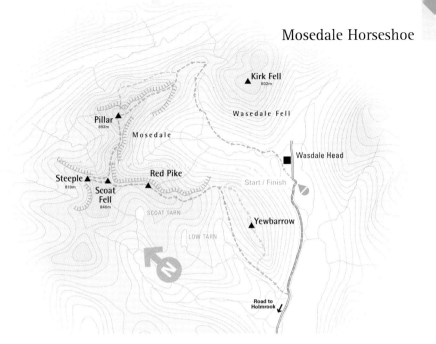

Walk #3 Mosedale Horseshoe

START	▶	WASDALE HEAD, 1865 0870
FINISH	●	CIRCULAR ROUTE
TIME	🕒	5 HOURS+
GRADE	🧭	NAVIGATION ● ● ● ●
	⛰	TERRAIN ● ● ●
	🌀	SEVERITY ● ● ● ● ●

This stunning walk is a shortened version of the full Mosedale horseshoe (omitting Kirk Fell on the ascent and Yewbarrow on the descent) taking in Looking Stead, Pillar, Little Scoat Fell, Steeple and Red Pike for a superb day out and a variety of geological features. The full horseshoe is a much bigger challenge and makes for a fairly long and tiring day without much time to spend on appreciating the surroundings. Yewbarrow

Mosedale Horseshoe

could be added at the end without much extra effort, but it does require a scramble on the ascent and an unpleasantly steep descent at the end of an already long day. The route I describe therefore omits this last mountain top, however much it cries out to be climbed for the sheer excellence of its 1.5km summit ridge walk.

The rocks seen on the walk are mainly from the Lower Borrowdale Volcanic Group: largely andesitic lavas and intrusions ('sills') interbedded with thin layers of sandstone and breccias. There's plenty of opportunity to study the types of rocks making up these early eruptions of the overall Lakeland eruptive cycle. There are also some outcrops of pyroclastic rocks to be seen.

The walk also ensures plenty of very fine views, especially towards the central and northern fells. The scenery of the Scafell massif, as seen on the second half of the walk, is stark and compelling.

The skyline is perhaps best described as gnarled, which my dictionary defines as: knotty, contorted, rugged and weather-beaten (which just about sums it up). The gnarly profile is the product of glacial carving and gouging of tough, resistant pyroclastic rocks. To our modern minds, the scene on a sunny day seems to us to be a wild, unmanaged, natural beauty, the antithesis of drive for order and predictability that defines the culture of our daily life. Thus the knotty, rugged, contorted landscape appeals to our aesthetic outlook.

We should not overlook the weather-beaten part of the definition for these fells are surely that. Indeed, I should add that the dictionary also offers two more characteristics to its definition of gnarled: ill-natured and bad-tempered. This side of the gnarly pyroclastic landscape of the central fells can certainly be displayed in tempestuous weather when the fells can be very unpleasant places indeed.

This walk is on open exposed fells to the west of Lakeland, facing the sea and prevailing winds, so feeling the full force of any ill-natured weather. This walk is best kept for good weather, not just because in bad weather you will miss the superb views it offers but also as these hills are particularly battered and beaten by westerly storms.

The walk gives an excellent overview of several glacial valleys. Mosedale is particularly striking, both on the climb to Black Sail Pass (by turning round and looking backwards every so often) and on the narrow ridge between Pillar and Little Scoat Fell (see photo 6.3). Plenty of glacial moraines can be seen, especially in Wasdale,

Mosedale and Ennerdale on the ascent. The valleys which lay below your feet on this high ridge walk are glacial in origin and once held large 'valley' glaciers which carved their present shape. (Note that the actual mountain sides are curtained behind screes in their lower reaches, so the valleys appear less steep-sided than they actually are.)

Photo w3.1 | Ennerdale seen from Black Sail Pass. Note the hummocky moraines in the valley floor and the parallel side valley (upper right to right centre) formed along the line of a parallel fault to the main fault running along the valley.

The walk starts from the car park at Wasdale Head, heading towards and through the complex of buildings around Wasdale Inn. Then aim for the base of Kirk Fell, picking up the footpath into Mosedale at about 1875 0915. After a couple of hundred metres, the views into Mosedale open up and you can see several substantial moraines. Indeed, you pass through some of them farther along the track.

Follow the track for another kilometre or so, bearing right at the path junction at about 1825 1020 when you start to ascend gradually towards Black Sail Pass. Bear sharp left at one point to cross Gatherstone Beck at a ford above some small waterfalls. Another 250m of ascent brings you out to Black Sail Pass where, if you head to a point where views open up, you look out on Ennerdale and the High Stile ridge and also towards Haystacks, Green Gable and Great Gable.

The display of hummocky moraines is quite stunning (see photo w3.1). The moraines are made up of an unsorted mass of sand and gravel, pebbles and boulders. They were deposited by glaciers moving down Ennerdale (from right to left in the photograph).

Mosedale Horseshoe

Photo w3.2 | View towards Little Scoat Fell.

Ennerdale has been created on the line of a major fault. However, just about opposite where you are standing (or sitting if you stop here for a break) the fault has split into two parallel legs. The main leg of the fault follows the main valley course up to the col between Green Gable and Great Gable (where its line is continued down Aaron Slack, past Sty Tarn and Sprinkling Tarn, up between Great End and Seathwaite Fell/Allen Crags, past Angle Tarn, down Rosset Gill and along Langdale and beyond). The subsidiary fault runs along the smaller valley containing Tongue Beck that you can see on the opposite side of the main valley. This fault continues up east of the summit of Green Gable then heads down, swinging right, to meet another fault coming in at right angles along Styhead Gill where it peters out. The minor fault is therefore the seat of a sort of parallel hanging valley above the main valley.

Return to the track up towards the summit of Pillar, passing many outcrops of mainly lavas of the Lower BVG. Around Looking Stead, you pass rhyolitic rocks (tuffs and 'reworked' sedimentary rocks) from one of the comparatively minor pyroclastic eruptions that occurred during the main lava eruptions phase early in the overall Lakeland eruptive cycle. There are magnificent views of Mosedale as you ascend.

Just beyond Looking Stead you can divert right to pick up the climbers' traverse, which takes you between the crags to just beyond Robinson's Crag and then steeply up to the summit. This is an utterly wonderful walk, but its steepness may not suit everyone especially at the start of a long day. The alternative is to continue up the gently rising main track until you reach the summit.

The rocks passed on the way are mainly lavas from the Lower BVG with only intermittent seams of rhyolitic rocks. The summit area is wide and rather flat, so a walk around the summit plateau is recommended for quality of views (especially for a view of Pillar Rock if you didn't come via the climbers' traverse).

Head towards Little Scoat Fell via Wind Gap (photo w3.2). As you climb up to Little Scoat Fell you cross out of the Lower BVG into an area of sedimentary tuffs and volcaniclastic sandstone (i.e. sedimentary rock made up of material eroded from nearby volcanic rocks and deposited under water). These outcrops make up the flat, rock-strewn, summit area. The pyroclastic rocks passed so far on the walk are all from minor intervals of pyroclastic eruptions that took place during the mainly effusive lava eruptions of the Lower BVG (later on, during the descent, we will pass some pyroclastic rocks from the Upper BVG).

Photo w3.3 |

View from near the summit of Little Scoat Fell towards Red Pike (right), the Scafell massif (centre) and Kirk Fell, left.

Mosedale Horseshoe

If you then head out to Steeple as is recommended, you very soon pass out of the volcaniclastic sandstone into an area of basaltic andesite between the summit area and Steeple,
which is yet another different rock type (andesitic lava). This rapid succession of rock types
accounts for the existence of the stuttered summit of Steeple. There are very fine views of
Ennerdale from Steeple (see photo 6.10).

Return to the summit of Little Scoat Fell, with its light-coloured rhyolitic rocks which
have been subject to 'periglacial' freeze-thaw action and now form a boulder field (photo 3.1).
From here start heading down towards Red Pike, once again in the rocks of the Lower BVG
formation (photo w3.4). When you start to rise again towards the summit of Red Pike, take
the smaller track along the edge of the crags over Mosedale. You pass some good examples of
andesite lava on descent to Dale Head.

At Dale Head you must decide upon a route of descent. The direct descent to Mosedale
via the Dorhead Screes is now highly unstable and use of the route causes erosion, shifting
the scree downwards. It is therefore best to take the longer but gentler descent via Over Beck
or the brutal, but rewarding in terms of views, scramble up and slither down Yewbarrow.

Photo w3.4 | View from Red Pike to Kirk Fell (far left), Great Gable (left), the
Scafell massif (centre and distant right) and Yewbarrow (centre right).

Photo w3.5 | View towards Scafell Pike (centre) and Scafell (right). Lingmell Gill follows the fault line that runs up to Mickledore, the deep-sided col between the two peaks.

The described route follows the easiest option, via Over Beck. The rocks are again mainly Lower BVG lavas and intrusive dark (andesites and basalts) rocks with the odd appearance of rhyolitic rocks, for example from about 1685 0830 to about 1670 0735.

These rhyolitic rocks represent the present-day surface outcrop of the earliest of the Upper BVG rocks. The outcrop is only the eroded top of a wedge that runs downwards from these pyroclastic rocks, surrounded by Lower BVG lavas. There is another surface outcrop of a wedge of Upper BVG pyroclastic rocks on the summit ridge of Yewbarrow. Either of these wedges may mark the western limit of the great Scafell caldera or they may be exposed plumbing conduits on the line of faults, leading to a fissure vent through which the magma was rising but cooled before eruption. The Upper BVG outcrop in Over Beck is bounded on the west by the Dorehead Fault which runs down to Wastwater. Here the fault is displaced to the west along the line of a major fault running along Wastwater. After a couple of hundred metres the Dorehead fault continues as a curving line that passes just west of the summit of Illgill Head. The main boundary between the Lower BVG and the Upper BVG (between lava landscapes and tuff terrain) lies on the western slopes of the Scafell massif.

Mosedale Horseshoe

The exact line of the caldera is not certain. The pyroclastic rocks around here and on Illgill Head are part of the 'outliers' of Upper BVG and thus represent the great shift from effusive to pyroclastic eruptions which characterises the two parts of the Lakeland volcanic episode. These questions are worth considering in relation to the landscape as you descend towards Over Beck, down through the valley, around the base of Yewbarrow and back to Wasdale Head. For example, there is a clear difference between the shape of Kirk Fell and Yewbarrow as you descend from Red Pike (see photo w3.4). As you get lower and new vistas open up, you may find it profitable to see if you can detect changes in rock types on Illgill Head and on the Scafell massif, marked roughly by the outbreaks of crags seen for example on either side of the head of Lingmell Gill (see photo w3.5).

The views open up towards Wastwater as you descend, especially when you reach the lower part of the south-western ridge of Yewbarrow. Follow this down until you meet the footpath bearing sharp left to head back towards Wasdale Head, reaching the minor road at about 170 070. Then head back to Wasdale Head along the road, enjoying super ever-changing views of the surrounding fells, especially to the Scafell massif and Great Gable.

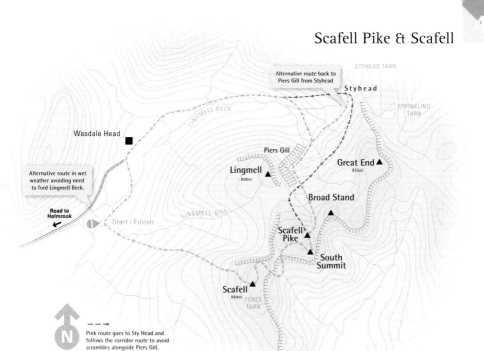

STYHEAD TARN

Alternative route back to
Piers Gill from Styhead

Styhead

SPRINKLING
TARN

LINGMELL BECK

Wasdale Head

Piers Gill

Great End ▲
910m

Lingmell ▲
806m

Alternative route in wet
weather avoiding need
to ford Lingmell Beck.

Road to
Holmrook

Start / Finish

LINGMELL GILL

Broad Stand
▲

Scafell
Pike ▲
▲
South
Summit

Scafell ▲
964m FOXES
TARN

N

Pink route goes to Sty Head and
follows the corridor route to avoid
scrambles alongside Piers Gill.

Walk #4 Scafell Pike & Scafell

START	▶	WASDALE HEAD (1865 0870) OR STY HEAD (2080 0940)
FINISH	●	CIRCULAR ROUTE
TIME	🕒	6 HOURS+
GRADE	🎧	NAVIGATION ● ● ●
	☁	TERRAIN ● ● ● ● ●
	⊗	SEVERITY ● ● ● ● ●

The Scafell massif is the product of a series of massive pyroclastic eruptions that occurred some 430 million years ago. The eruptions pumped out so much pyroclastic material that the land above the magma chamber collapsed in on itself, resulting in a large 'caldera' or roughly circular depression (to the east of Scafell Pike) which subsequently became filled with rain water. The extremely tough rocks produced by

Scafell Pike & Scafell

the eruptions have survived aeons of erosion so that now Scafell Pike is England's highest mountain.

By the term 'Scafell massif' I am referring to Scafell Pike and Scafell together with the surrounding peaks, Lingmell, Great End, Broad Crag and Ill Crag. These fells are part of the mountain summit area draped around the northern reaches of the Eskdale and Lingcove Beck valleys (Scafell and Scafell Pike, Esk Pike, Bowfell and Crinkle Crags). This group really forms a single mountain mass, which in turn is part of the central mountainous core of the Lakeland fells which includes Great Gable and Glaramara. This whole area has a similar geological history and structure, although with a widely varied range of rock types from different eruptions. Essentially this is the area created by the eruptions which also created the Scafell caldera. The caldera or 'fault-bounded depression' would soon have become filled with rain water and with sediment from subsequent eruptions and from 're-worked' material (eroded from rocks above lake level and carried down by streams).

This walk follows some of the most spectacular scenery and geological features of Scafell Pike and Scafell, largely avoiding the busiest tourist tracks. However, on the roof of England, it is inevitable that you will bump into many people. Some do unusual things such as carrying a refrigerator up to the summit and back. Others may be participating in one of the challenges involving many unhappy hours in a car or minibus driving between Ben Nevis, Scafell Pike and Snowdon in order to climb all three peaks within a twenty-four hour period. It is unlikely that those who do this sort of thing can afford to spare the time or effort needed to enjoy the scenery or the rocks.

This walk presents an alternative type of challenge, one which involves no driving but does require a lot of climbing and descending through stunning craggy scenery. It also offers the walker plenty of opportunities to take time to appreciate the various types of rock and array of geological and glacial features that can be seen by those who take the effort to look for them, and of course the wonderful views. Such features are in abundance on this stark and gnarly mountain mass, so it seems a shame to pass them by without a glance let alone a nod of acknowledgement.

The walk as described is pretty long and involves quite a lot of ascent. The climb up Scafell Pike involves about 900m of upwards slog and the ascent of Scafell adds

166

another 300m. As the trip out to Scafell involves a considerable climb and a long hard descent, an option to leave this section out is described. The cause of the problem is the deep col Mickledore between Scafell Pike and Scafell. This is centred on a fault and has been deepened by glacial action, leaving a steep rock wall on the Scafell side (Broad Stand). Unfortunately this stepped rock face is too steep for walkers, however inviting it may look. An additional 150m therefore has to be descended after getting down to Mickledore, to find a gully (another fault-determined feature) and a way up onto Scafell's summit. However, the extra effort is well worth it if you have the time and energy for the views (for example Scafell offers some of the best views of Scafell Pike) as well as the wealth of features to be seen.

The walk takes you through some of the most dramatic scenery created by the volcano-era eruptions and subsequent erosion and, as described, starts from Wasdale Head. It can also be started from Sty Head for those coming from Borrowdale or Honister, when leaving out the extra trip to Scafell is recommended as it takes you rather a long way from your destination.

From Wasdale Head follow the public footpath towards and beyond Burnthwaite farm. At about 202 093, where a path diverges left up towards Sty Head, keep right to follow the path beside Mires Beck for another 600m to about 208 093 where a path heading up left to Sty Head is also to be ignored. (However, if you prefer to miss out on the slightly difficult scramble up besides Piers Gill, you should make your way to Sty Head on one of several tracks. From there, pick up the Corridor route track, rejoining the main route at the top of Piers Gill.)

Continue alongside the stream for another 400m until just after the beck, coming down from Piers Gill, joins Mires Beck. Cross the stream and head uphill, above the Piers Gill beck, until after another 200m when you should take the right-hand split of the track.

After about another 350m, another stream has to be crossed at the base of Piers Gill. (This point can also be reached from Sty Head by following the Corridor Route until a narrow track branches off right towards Piers Gill.)

The view up into Piers Gill, after crossing the stream, is highly impressive and worth lingering over (see photo w4.1). You may wish to explore a short distance up the Gill, but it

Scafell Pike & Scafell

Photo w4.1

View up into the
lower section of
Piers Gill carved
along the line of a
dolerite intrusion.

is very steep and soon becomes impassable except to skilled and equipped climbers and you would have to re-descend the steep slope in order to continue.

Piers Gill is one of the main 'blackspots' for the local mountain rescue team. Care needs to be taken to follow the correct route and on the scramble mentioned below. Do not attempt to go up the Gill itself or to attempt any route on the right-hand side of the Gill (as you ascend). However, do not be put off this route if you have some hillwalking experience. Most of the incidents that lead to a call-out of the mountain rescue volunteers occur on descent where the scramble is harder (and also where people try to descend on the wrong side – the left-hand side – coming down). The main danger on the described route is the scramble (see below), but this is far less problematic than trying to descend here.

Piers Gill is one of several deeply engraved streams on this face of Scafell Pike (these are particularly well viewed from Walk 5 on Great Gable; see photo 6.12). These canyons were probably created by powerful streams running underneath glaciers which covered this area during the ice age. The streams would have been under considerable pressure and would also have carried large boulders, thus having enough power to carve these gorges out of weaknesses in the rocks.

This first deeply entrenched section of Piers Gill is cut out along a weakness created by a boundary between the dacitic lava of much of the rock outcrops around here (and on Scafell) and an intrusion of dolerite (one part of some complex sets of dolerite and andesitic intrusions in the area). However, where Piers Gill does a 90 degree turn further up, it turns away from the rock boundary and instead lies on a weakness created by a deep running fault. The other deeply cut gills in the area also exploit either faults or boundaries with intrusive rocks (see photo 6.12).

The track to be followed is on the left-hand side of the gill on the high ground above Piers Gill, initially close to the edge of the Gill, but with restricted views down into it. Take very great care if you get closer to the edge for better views, as the sides of the Gill are fairly precipitous.

Half-way up this first leg of the Gill, the track starts to swing left away from the Gill in order to cross a side gill coming down from Middleboot Knotts. The track here gets quite steep and involves some scrambling up a 10m high rock wall and care is needed. Keep a close eye out for the track as it tends to disappear from under foot only to re-appear a bit further on.

The scramble brings you out near the sharp 90 degree bend in the Piers Gill, where the chasm leaves the dolerite intrusion to follow a fault. You can leave the track here to stand not too near the edge (and with very great care) to peer down into the chasm, before returning to the track to continue the ascent.

The rocks are initially part of the 'Scafell Dacite' formation, which was erupted onto the surface to form a dome with sloping sides down which the lava flowed. On the crags to the left, Criscliffe Crags, geologists have measured the angle of the original slope of the side of the dome at 25 degrees. You will see more Scafell Dacite later on when you get to the summit plateau of Scafell.

Later, you may notice the bedding in the rocks you pass along this section of the walk (see Photo 3.7). These are sedimentary rocks within the Seathwaite Fell formation which were deposited within the Scafell Caldera.

Eventually you meet the Corridor Route (and probably the crowds too) onto which you must turn right for a short section to get under some crags. Then turn left immediately,

Scafell Pike & Scafell

Photo w4.2 | View from north-western ridge of Broad Crag towards
Broad Crag (right), Sty Head (centre) and Great Gable (left).

leaving the Corridor route to head up the continuation of Piers Gill between Broad Crag on the left and Dropping Crag on the right. There is a very good view from here of the chasm alongside which you have just ascended (see photo 4.5).

Continue up the gill towards the col between Broad Crag and Scafell Pike, following the line of the fault running up through Piers Gill. After about 250m, the crags on the left come down to near the track level. It is recommended that you take a small diversion to the left here to the skyline (at about 2155 0770) for excellent views down to Sty Head (photo w4.2).

Return to the track and continue upwards, with some steep scree to be negotiated for a few metres just before reaching the col (at about 2170 0745) and yet more magnificent views. Look back down the way you have come from to see the fault line of the valley and its extension into Piers Gill. Bear right for the final 100m of ascent to the summit of Scafell Pike.

The summit cairn is at the north-western end of the wide stony summit plateau that dips gently southwards. The wide plateau limits the full extent of the views from the summit and the summit cairn area is always extremely busy, so it makes sense to head towards the south summit (at about 2160 0710) which is much quieter, has much better views and provides a good place to sit.

Keep an eye on the rock outcrops on the left as you approach the south summit. Indeed, a short diversion to study these outcrops is recommended, as you can see bedding and slumping features in the rocks around here (photos w4.3, w4.4 and w4.5).

From the south summit in clear weather you can see a panorama of peaks running from Ill Crag on your immediate left (with tilted bedding easily seen in the cliff face), to Esk Pike, Bowfell (straight ahead) and Crinkle Crags. Below lies the twisting valley of Eskdale with its massive rock steps (see photo 6.13). You also have a fine view here of Scafell on the right.

The largely pyroclastic eruptions which created the rocks of the Scafell massif were among the first of the Upper BVG eruptions. They took place following the extensive early lava eruptions to the north and west, which created the Lower BVG rocks. The first eruptions of pyroclasts of the Whorneyside Formation took place from vents to the north-west of the Scafell massif on the south-west side of Buttermere. Material covering the area between Great Gable and Eagle Crags south- and east-wards in a curving arc (with further outcrops farther south) was pumped out. These eruptions produced welded tuffs and other tuffs including air-fall tuffs.

Photo w4.3, w4.4 & w4.5
Sedimentary structures in Seathwaite
Fells formation rocks near the
south summit of Scafell Pike.

Scafell Pike & Scafell

Immediately after, another set of pyroclastic eruptions occurred producing the Airy's Bridge Formation (mainly rhyolitic tuffs and welded tuffs). This was a massive and closely timed set of eruptions, emitting about 150 million cubic metres of material and largely voiding the underground magma chamber. As a result, the domed area of land crashed down along the lines of faults that had also served as fissure vents. This created the Scafell caldera, which then filled with water.

The landslips were followed by sliding and slumping of the material that had been erupted, confusing the geological evidence. The landslips also allowed water to come into contact with the emerging magma, adding explosive power to the eruptions. The caldera floor sagged as the eruptions tore at its foundations, opening new fissure-vents for further material to be erupted. Further eruptions of pyroclastic material and lava, known as the Scafell Dacite (passed earlier in the Piers Gill area and also to be encountered later on the summit plateau of Scafell), marked the end of the episode. Material eroded from the surrounding rocks was then deposited in the caldera lake(s), eventually forming sedimentary rocks.

Head back towards the summit of Scafell Pike then, if carrying on towards Scafell, bear left to pick up one of the tracks down to Mickledore, the narrow col between Scafell Pike and

Photo w4.6 | Looking towards Scafell and Mickledore (centre right). The route to the summit of Scafell involves a descent to Mickledore, then left down the edge of the crags to the obvious vertical line in the centre of the photo which reaches into the corrie holding Foxes Tarn. The route then bears diagonally to the right, heading up to the summit plateau of Scafell.

Photo w4.7 (right) | Deep Gill from near Symonds Knott, Scafell.

Photo w4.8 (above) | View south-east from the summit of Scafell.

Scafell (see Photo w4.6) with an interesting variety of rocks passing underfoot. If you don't wish to continue to Scafell, then follow the main tourist track from the summit of Scafell Pike to Lingmoor Col and then bear initially south-west down Brown Tongue (or if returning to Sty Head, descend using the Corridor route).

Otherwise continue to Mickledore. However, do take time to stop and study the north-eastern face of Scafell as you drop down towards the col, to appreciate the vast, brutal grandeur of the rock and also to identify the route up to the summit plateau of Scafell. The route is up the obvious gully some way to the left of the col, which means that you have another 150m of descent even after you have reached the col (see photo w4.6).

The route up the trench-like gully (created on a boundary between rock types) is steep, slippery in places and requires one or two short scrambles, but is highly atmospheric. Breccias can be seen in the outcrops. Eventually you are debouched unexpectedly into the corrie holding the tiny Foxes Tarn. This is an enchanting but usually sunless spot, more attractive in the diffuse light of cloudy days than on sunny ones when the corrie is at least partly in the shade. Continue up the steep scree track at the north of the tarn. For some reason most people struggle up the slippery scree, ignoring the stretches of stone track just to the left. If you keep an eye out for these you will have a much easier time. The track is initially steep but the angle soon relents and you arrive on the summit plateau.

Rather than heading straight for the summit (on the left), bear right towards the slight eminence of Symonds Knott (a good spot for finding a sheltered place to sit, eat and drink as well as ponder the views). Then head round to the V-shaped clench at the top of Deep Gill with its vertiginous views down into the valley below (photo w4.7) and then to the rocks to the right (above Mickledore) for the best views towards Scafell Pike. Indeed, this is the best position for appreciating the structure of its summit area.

All these rocks around this immediate area are part of the Scafell Dacite, as also seen earlier on. This was erupted late in the cycle that created the Scafell caldera. It is at its thickest in the Dow Crag (in the Coniston area) and Littlearrow Cove area (east of Scafell Pike), where the vent was probably located and where large amounts of lava poured out creating a classic cone-shaped volcano. Breccia can be seen in some places, but not everywhere. However, flow-banding is a common feature of the Scafell Dacite.

When you've seen enough head towards the summit cairn and then walk for 150 metres or so along the summit ridge for views towards the southern slopes of Scafell and Eskdale.

Return to the summit and carry on until there is a drop of 5–10m, where you pick up a track heading left (north-west) towards Wasdale Head. This is an altogether pretty unpleasant descent with steep, slippery stony tracks at the beginning and again later on (when dropping down near Rakehead Crag). It is the most direct walkers' route at the end of a long day. (However, it is not the track shown on the OS map by black dashes, which heads to Green How for a longer but somewhat easier descent).

After the first steep stony section of descent you meet a grassy area. Just after meeting a little stream you come across some rock outcrops including welded tuffs. This is an isolated 'outlier' of Upper BVG rocks in the lavas of the Lower BVG. The pyroclastic Upper BVG rocks underlie the slight prominence of Green How. The boundary of the Scafell caldera may well lie here or slightly higher.

The track then heads towards the head of the crags overlooking Lingmell Gill and then suddenly turns right over the edge of the ridgeline to drop down another unpleasant, even steeper and more slithery, track towards the intake fence. A final steep descent on grass is necessary before your knees get a respite as you approach the footbridge over Lingmell Gill, from where footpaths lead you back to Wasdale Head.

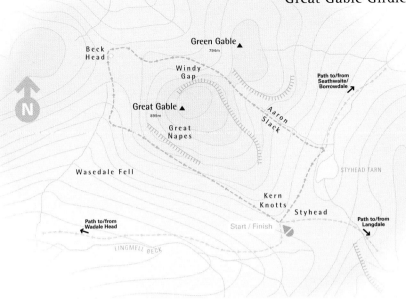

Walk #5 Great Gable Girdle

START	▶	STY HEAD (217 094)
FINISH	●	CIRCULAR ROUTE
TIME	◕	5 HOURS+
GRADE	⊙	NAVIGATION ● ● ●
	◔	TERRAIN ● ● ● ●
	⊗	SEVERITY ● ● ● ●

Great Gable is one of the most popular peaks in the western fells, its pudding-like summit area poking out above the immediately surrounding peaks making it highly recognisable from afar. It also offers marvellous views if you walk round the summit plateau. However, its very height also means that its topmost reaches are quite often shrouded in cloud.

Great Gable Girdle

This walk, which uses some old climbers' tracks suspended improbably in the middle of the great faces of the mountain, is a fantastic walk in itself. It really comes into its own when the summit is hidden in mist, providing great views over to the Scafell massif, along Wastwater and finally down Ennerdale from its very head. For the hillwalker interested in geology and scenery, the walk also brings you close to lots of fascinating rocks.

The recommended route takes you on a series of balcony walks high up on the sides of glacial valleys. Wastwater is a reminder of the work done by the glaciers in cutting down into the valley floor: the bottom of Wastwater is below sea level. The glacier would have filled the entire valley. You can also clearly see moraines left by the retreating glacier and some of the scores of thousands of boulders deposited by glacial meltwater in the valley and now largely collected and piled into stonewalls.

The walk is largely along the boundary between the exposed volcanic crags and the top of the mass of scree girdling the lower slopes. It perfectly illustrates two of the most important influences in the shaping of Lakeland: volcanic rocks and glaciation.

The section of the 'girdle' between Sty Head and the south-east ridge of Great Gable overlooking Gable Beck is a sensational experience, with some minor scrambling and much scrabbling over boulders small, large and gigantic. The narrow track traverses below the great crags of Kern Knotts and Great Napes, passing Napes Needle and the Sphinx Rock. The crags and scree slopes attract your attention upwards, while the picturesque vista towards Wastwater fights to draw your gaze downwards.

I have started my recommended route at Sty Head so that the walk can be approached from Wasdale Head or Borrowdale (or indeed from Honister, with the slight variation of starting and finishing at the head of Aaron Slack rather than Sty Head).

Start from the large cairn and nearby stretcher box, heading on a faint track on a bearing of 250 degrees. The track soon disappears as you cross a small boggy area, then re-appears over the crest of a low outcrop ahead of you. Once on this small outcrop you can see the track in front of you heading towards bigger crags, Kern Knotts, on the looming south face of Great Gable (see photo w5.1).

When the track reaches the crags, scramble over some of the big boulders and outcrops to pick up the track again on the other side. Take the opportunity around here to look up at

the rocks above you. These are lavas of the Scafell Dacite formation (which occurred between episodes of pyroclastic eruptions of the Upper Borrowdale Volcanic Group sub-cycle). Breccia, flow-banding and slumping features can be seen in the rocks here and for a short distance further (see photos w5.2, w5.3 and w5.4).

The track splits into two slightly further on. It is probably best to take the higher track, but it doesn't really matter which you follow as they re-unite after a while, and you then cross a small fault carpeted with scree. This is roughly opposite the lowest part of Piers Gill (Walk

Photo w5.2 (left) | Slumping in flow-banding.

Photo w5.3 (right) | Breccia.

Great Gable Girdle

Photo w5.4

Disturbed, partly
brecciated, flow-banding.

4) and is a good spot to study this and other gills deeply etched into the lower slopes of the Scafell massif (see photo 6.12). You also get a series of super views down into Wasdale and on to Wastwater (see photo 6.4).

From here on the steadily rising route requires you to cross a series of rock bastions or buttresses separated by scree slopes. The track roughly follows the boundary between the dacite lavas above and right and the pyroclastic tuffs below and left. From around here you see a greater variety of rock, including coarse tuffs and breccias.

Eventually you come across Napes Needle and then the Sphinx Rock, both a short distance higher up the slope. You may want to visit them for a closer look or to take photos, but remember that you will need to descend to resume the walk along the track.

Shortly after Napes Needle you meet the one difficult section on the walk. First it is necessary to edge gingerly around (or to climb over) an awkward rock. There is then is a 2 metre climb needed down another rock where great care is essential. Either identify footholds from above and turn in to face the rock to descend or inelegantly edge down, from a sitting position, over the grassy left edge of the rock.

The track soon brings you to the south-western ridge of Great Gable, where a track from the summit to Wasdale crosses your route (it is possible to descend from here, but it would be far easier to carry on to Beck Head and head down from there for a less steep and slippery route).

Here you are roughly above the edge of the great Scafell caldera. Its western edge perhaps crosses roughly north up the side of the Scafell massif, just above Black Crag, below Lingmell and across to the slopes of Great Gable about 50m below you. It continues round towards Beck Head then turns north-east round the north-western slopes of Great Gable, Green Gable, Base Brown and Thorneythwaite Fell.

To the south, on the slopes of Lingmell and Scafell, you can also see the steeper, craggier shape of the slope at the margins of the caldera. However, do remember that there has been a re-ordering of the original relative heights; when it was created the caldera would have been lower than the lava rocks to the west. After millions of years of uplift and erosion, the positions are reversed and the tougher pyroclastic rocks of the upper reaches of the Scafell massif now stand proud and high.

To carry on to Beck Head, follow the track north from the ridge line with a completely new vista opening out. It's also worth taking a bit of time along here to study the moraines, fences and piles of boulders in the valley below. The track on this section of the walk is largely level, crossing blocky scree. The track dips slightly about 100m from the col to carry on along some grass below the scree.

From the col, head right to go up to the base of the north-western ridge of Great Gable. Head up the ridge itself for about 50m of ascent until you come across two small cairns on the left side of the track (at about 2987 1060), as you approach the base of the crags above and to the left. (Alternatively, you could carry on up the ridge to the summit of Great Gable if you prefer.)

Bear left between the two cairns to pick up a small track which guides you across the third face of Great Gable, its north face to Windy Gap at the head of Aaron Slack. This is usually a much darker place than the other faces, except when cloud ensures equal illumination all round the mountain.

The track ascends gently across the sloping face at the top of the scree, slowly getting closer to the base of the crags where more of the Scafell Dacite lava crops out. Near the end, the track drops down slightly as you approach the crags then resumes its ascent to bring you out at Windy Gap.

If the weather is clear and you want to visit the summit of Great Gable (Scafell Dacite lava) you can ascend from here, though you will need to re-descend to this point to follow the fourth leg of the walk: the descent of Aaron Slack.

Aaron Slack, which sharply separates Great Gable from its neighbour Green Gable, lies on the line of one of the major faults in Lakeland. It runs through Ennerdale, Aaron Slack, past Styhead Tarn, Sprinkling Tarn, Angle Tarn, Rosset Gill and Langdale and Lingmoor Fell. As you descend you can pick out the line of fault ahead of you, including a small gorge which runs parallel to the base of crags to its right and across the head of a valley (Ruddy Gill) to its left. The fault line is not entirely a straight line, there being something of a displacement around Sty Head, but this is often the case with faults which tend to curve.

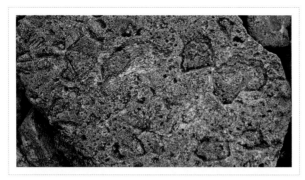

Photo w5.5
Brecciated lava.

Photo w5.6
Flow-banding in lava
as seen in a slab in the
track near Sty Head.

The descent of Aaron Slack is not very steep, but is very slithery on the mass of scree underfoot. You will need to stop and sit down if you want to study the fault line and other features. Take time to look at the slabs of rock used in the improved sections of the track; you can see some fascinating breccias (see photo w5.5). This is the fourth face of Great Gable, but unlike the others this one feels much more like a mountain descent rather than a traverse. It is nonetheless an excellent part of a great walk, and eventually brings you back to Sty Head.

If returning to Wasdale Head, I recommend following the track (shown by black dashes on the OS map) that descends into Spout Head Gill, sticking close to the stream.

Start by following the main stone path towards the Corridor route. You should see on the path a stunning example of flow-banding in lava as a block in the path (see photo w5.6). Shortly after that, pass some pinkish rocks on the left then pick up the path leading below some crags on the left into the upper bowl at the head of the valley. Note the excellent views up into the deep gills above on the left (and later on good views of Piers Gill). Devise your own route down to the main valley, ensuring that you don't leave it too late to get onto the right bank of the stream to pick up the bridleway to Branthwaite Farm.

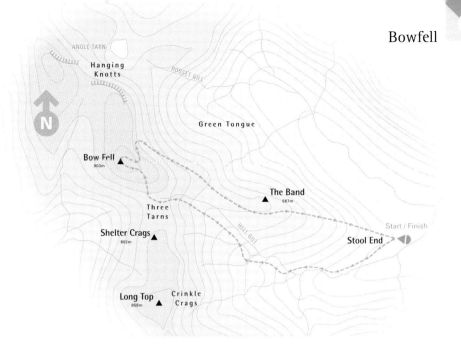

Walk #6 Bowfell

START	▶	BUS STOP/CAR PARK AT OLD DUNGEON GHYLL (285 059)
FINISH	●	CIRCULAR ROUTE
TIME	◔	5 HOURS+
GRADE	⊕	NAVIGATION ● ● ● ○
	☁	TERRAIN ● ● ● ●
	⊗	SEVERITY ● ● ● ● ●

Bowfell is undoubtedly one of Lakeland's finest mountains with utterly wonderful views of the Scafell massif, the central fells, Langdale and the fells to the south. For the lover of geology and scenery, it also has an immensely interesting summit area. This walk follows a rather adventurous route up to Bowfell using an old climbers' track and then descends alongside the appropriately named Hell's Gill.

Bowfell

It is fairly common to include a walk across Crinkle Crags or over Esk Pike and Rosset Pike in the traverse of Bowfell, but with so many interesting rocks and other features it pays to take your time and get the most out of a round of just Bowfell. It also makes sense to do it in good clear weather as the views from Bowfell's summit are an important part of the walk.

Bowfell is a rocky mountain (see Photo 5.8). Its summit plateau is a mass of boulders and tors with numerous fascinating rock features to be seen, especially the variety of bedding in sedimentary rock and other features such as disturbed or slumped bedding. The uppermost part of the mountain is made of volcanic tuffs and breccias. Between the top summit area and the col of Three Tarns, however, the rocks are by and large sedimentary rocks. These sedimentary tuffs and 'volcaniclastic sandstones' are made up of particles eroded from volcanic rock on the edges of a caldera and deposited by streams into a depression in a collapsed caldera.

The material settled to become hardened into rock. However, the still relatively soft material was at times shaken by earthquakes as further tectonic activity took place nearby. Slumps where the beds have been disturbed can be seen in many places on the descent on the described walk from the summit area to Three Tarns.

The later effect of earth movements is also brought home very effectively by the absolutely clear tilt of the beds of sandstone, especially well exposed in the 'Great Slab' passed on the described ascent. This illustrates the extent to which some of beds of rocks were tilted and folded during the mountain-building episode that followed the volcanic cycle.

There are plenty of opportunities to extend the walk to Esk Pike or Crinkle Crags. However, I describe a route that returns via Three Tarns and then down towards an impressive glacial meltwater channel, Hell Gill and a typical Lakeland waterfall, Whorneyside Force. This allows for an unhurried walk with ample time for looking closely at the amazing variety of rocks to be seen.

Take the footpath starting from where the road bends up towards Blea Tarn heading to Stool Farm. Shortly after passing through the farmyard, bear right up the signposted track via The Band. The track goes round to the left of the crags, then heads steeply upwards before again heading left to avoid crags near the top of The Band. The path is well made, but suffers

from a surfeit of cairns. Eventually you reach an open area on the ridge top without any higher crags to your right. Carry on, passing two cairns close together on the right of the path and one on the left at a slight depression, then follow a side track off to the right (at about 2540 0615). The track initially crosses a boggy area, then starts to rise and becomes a well-made path.

The track rises steadily to meet the edge of the crags at about 2499 0634, overlooking the head of Mickleden and Rosset Gill. A narrow track leaves from here to cross Flat Crags, roughly where the crags end and the scree slope begins. The undulating track crossing the cliff face is occasionally vertiginous, sometimes washed away and every now and then presents mildly challenging sections to be overcome (see Photo w6.1). If you don't wish to use this track, instead stick with the track from The Band to Three Tarns, ascending there and re-joining the main route at the top of the great sloping slab met on the right of the main path just below the summit of Bowfell.

The views and the high balcony-like position of the traverse are spectacular, especially looking down into the valley below. This great glacial valley was carved into an area of weakness created by a major fault running down from Ennerdale and below Great Gable to Rosset Gill, then via Mickleden, passing near Side Pike and Blea Tarn (where it runs up a deeply etched and wooded gully). A subsidiary fault leaves the main fault near the top of Rosset Gill then runs through the base of the crags below you, up and over the flat area where you left the main track, down Buscoe and through Hell Gill, beyond which it runs out on meeting another fault coming at right angles down Crinkle Gill.

The complex of moraines left by the retreating glacier is visible in the valley below.

Initially the rocks are not very distinctive, but after a hundred metres or so you can see

Bowfell

Photo w6.2 | The Great Slab with more exposed beds to its right, looking down from near the summit of Bowfell. Seen in the distance are Langdale Pikes (left), Great Langdale (centre) and Lingmoor Fell (right).

Photo w6.3 | A section through the Great Slab alongside the described route, showing various beds.

very clear signs of bedding in the rocks upon looking up. Eventually the track takes you across to Bowfell Buttress, where it turns upwards alongside a steep bouldery slope with crags to the right and the Great Slab of rock on the left, finally delivering you to the summit plateau area at the top of the Great Slab (see Photos w6.2 and w6.3).

You will see lots of bedding, ripples and some slumping in the rocks around here as you ascend to the summit and later on during the descent towards Three Tarns. These rocks are air-fall tuffs and volcaniclastic sandstones that settled into the lake created when water filled the Scafell Caldera. They therefore come from eruptions during the period after the main caldera-creating eruptive episode. The slumping is evidence that there were still a lot of earthquakes in this later episode.

From the Great Slab, head up the main track to the summit tor. As you get closer to the summit itself these bedding features disappear, although cleavage can be seen. However, your attention will inevitably be drawn away from the rocks to the stunning vista which opens up over upper Eskdale and towards the Scafell massif and the rippling felltops beyond.

When you are ready to descend, avoid the section of the main track you used for the final part of the ascent and head down from the western side of the summit. When necessary, bear east and rejoin the main track for the descent of the summit plateau to Three Tarns. From below the summit to about half-way down to Three Tarns, you will see a superb set of outcrops and boulders displaying bedding and slumping features. These really are well worth leaving the track to study in detail every now and then, to see the variety of slumping in the beds.

At the col take the track descending into Buscoe and on to the head of Hell Gill (see Photo w6.8), taking care not to get too close to the edges of this deep chasm that was cut by a powerful, rock-laden stream running at great pressure underneath a glacier. The gill was cut on the line of a fault, a subsidiary fault running across under Bowfell (as seen earlier on) from the major Rosset Gill fault.

Photo W6.4 & w6.5 | Slumping in beds of sedimentary rock caused by earthquakes dislodging partially hardened sediments (see also Photo 3.11).

Photo w6.6 | Ripples left by water currents on the top of a bed of deposits that eventually became sedimentary rock.

Photo w6.7 | Structures in sedimentary tuffs.

Bowfell

Photo w6.8 |

The overall tilt of the rocks caused by mountain-building forces during continental collision.

Follow the track back down towards where the streams from Hell Gill, Crinkle Gill, Isaac Gill and Browney Gill converge. Shortly after, the stream cuts through a small gorge at a knoll (at about 2653 0526) carved on the junction of Lower Borrowdale Volcanic Group lavas and later Upper BVG pyroclastic rocks. Follow the track back to Stool End farm and back to the finish point.

Note the wide stream bed and the mass of rounded boulders it contains. Note also the rounded boulders making up the many stone walls and, in places, the way the boulders have been used to line a deepened stream bed. Not long ago, the entire valley floor would have been covered by boulders. There has clearly been an enormous amount of effort put into clearing the ground to make the green fields we see today.

Photo w6.9 |

The head of Hell Gill, with Pike o'Blisco in the centre background.

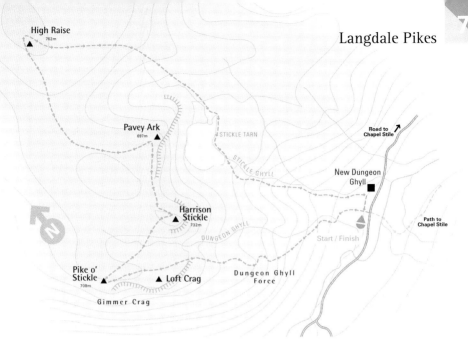

Walk #7 Langdale Pikes

START	▶	BUS STOP/CAR PARK SOUTH OF NEW DUNGEON GHYLL HOTEL (294 063)
FINISH	●	CIRCULAR ROUTE
TIME	☕	6 HOURS+
GRADE	🎧	NAVIGATION ● ● ● ●
	☁	TERRAIN ● ● ● ●
	⊗	SEVERITY ● ● ●

The Langdale Pikes, as seen from Great Langdale, form one of Lakeland's iconic views. As you head up the valley towards the Pikes from the east, both Harrison Stickle and Loft Crag appear to be pointed, rocky mountain tops, with steep craggy buttresses running down to the valley, riven by deep gills. The crags of Pavey Ark run off to the right from behind Harrison Stickle, a crest of rock adding a hint of further drama to the overall view (see Photo 5.9).

Langdale Pikes

These picture-book mountains have become favourite spots for fellwalkers and other visitors to Lakeland in search of sublime natural beauty.

The summits look their most impressive from the valley floor, where the Pikes rear up looking like impressive mountain peaks. Climbing up the steep mountain tracks, usually alongside or near one of the gills, the impression is enforced of being engaged in an ascent to a pointed summit. On nearing the tops the illusion is rudely squashed. The scene that comes into view is of an immense glacial plateau rather than the expected drops on the other side of the peaks (see Photo 5.10). From the Pikes' highest points one's attention is drawn ineluctably back towards Great Langdale and the mountain ranges arranged in an arc at the head of the glacial valley.

The plateau to the north has almost the opposite effect, its gently rolling grassy slopes pushing one's eyes away towards the more ravished and attention-grabbing landscapes of the peaks to the south and west. This plateau, running north to High Raise and beyond, sits right in the geographical centre of Lakeland – bang in the middle of the area affected by the powerful pyroclastic eruptions that occurred during and after the creation of the Scafell caldera, and at the core of the ice fields that settled on Lakeland many millions of years later. The plateau sits just about where the Lakeland ice cap would have been at its thickest.

While tens of thousands trek up to the Pikes, visit some of the various high points and then stumble back down again, far fewer venture north across these rolling grassy landscapes. This is a pity, for here you gain a more complete feel for the landscape of the mountain area. The Pikes, however conspicuously attractive they look from below, only actually form the marginal feature of a barely noticeable raised rim to the edge of the plateau. While High Raise may not appear much of a summit from the Pikes, it is in fact a very fine viewpoint and well worth a visit by fellwalkers.

This walk takes you up to the Langdale Pikes but does so via Sergeant Man, High Raise and the rolling grassy landscape before heading to the Pikes themselves to give a rounded view of the mountainscape. Here you can try to imagine a kilometre of ice sitting above you, pressing down on the mountains, scraping away the rock in places and sliding over it in others to leave this rocky-boggy wilderness. It then becomes easier to conceive of the way ice carved the other features seen on this walk, such as the exposed rock summits of the Pikes themselves, the deep glacial trench of Great

Photo w7.1 | View towards Stickle Tarn from near Sergeant Man. Rock outcrops visible on Pavey Ark (right) also underlie the hummocky terrain to the left of Stickle Tarn (just visible in the centre of the picture).

Langdale, the ice-breaching points (for example, between Lingmoor Fell and Pike o' Blisco or Langdale Comb), the glacial corrie and lake of Stickle Tarn and the glacial meltwater channels of Dungeon Ghyll. (Note that Ghyll is an old spelling of gill, which survives only in a few names such as Stickle Ghyll). The walk also brings you up close to a variety of pyroclastic and sedimentary rocks, including some very unusual breccias on Pavey Ark.

The walk can be cut short at a number of points, so it is not necessary to go to all the peaks as described if you don't want such a long walk. If you do shorten the walk, I advise making sure that you don't leave out Pavey Ark, for here you will see some of the most amazing volcanic rocks in the district. The walk ends with a descent alongside the aptly named Dungeon Ghyll.

Start from the notice board in the National Trust car park next to the Old Dungeon Ghyll Hotel, following the track for Stickle Ghyll. Cross to the right-hand side of the stream and ascend steeply up a well-made track clambering up over a variety of volcanic rocks, mainly tuffs with some breccias and lava. Higher up, the track crosses the stream to finish its journey

Langdale Pikes

on the other side, but if the stream is in spate you can continue on the same side if you prefer. In either case, you soon come out at the outlet from Stickle Tarn, cupped in a great hollow carved into the crags of Pavey Ark and Harrison Stickle.

You may well want to stop here for a break to admire the immense crags of Pavey Ark with the sharp line of Jack's Rake slanting upwards across the cliff face. This is a scrambling route, and requires something of a head for heights and scrambling experience. Our much more gentle route bears right on reaching the outlet following the lake shore, at first heading roughly northwards. The terrain ahead and to the right appears to be a mess of wholly uncoordinated hummocks and dips but, as will be seen on the way up to Sergeant Man, there is a pattern to this very confusing landscape.

Continue on the track of the public footpath beyond the end of the lake heading into Bright Beck, continuing until about 288 081 where an unmarked track follows roughly alongside a stream descending from just east of Sergeant Man. Follow the stream, stopping now and then to turn around to appreciate the magnificent views, and then head for the summit. When you turn to look at the landscape, you can now see that the hummocky landscape below is formed from ribs of rock running roughly parallel to your view. The hummocky terrain is a product of the way glaciers have ridden over this landscape, carving away at weaknesses exposed where the rock ribs intersect the surface. These rock ribs are mainly made up of the same tough volcanic rock that forms Pavey Ark (see Photo w7.1).

You pass some outcrops of the same rocks as you climb up near the stream, examples of coarse tuffs and breccias (see Photo w7.2). Looking towards Pavey Ark from near Sergeant Man, you can see the rocks breaking out at the surface above Bright Beck which are, if you could see them close up, the same breccia. (More of this will be seen later on the summit of Pavey Ark.) Beyond these obvious rock outcrops, the terrain is smoothly rounded and grassy and this is formed by somewhat softer underlying sedimentary and volcanic rocks. You can see some sedimentary rocks close up on the outrops in the summit area of Sergeant Man (see Photo w7.3). The softer sedimentary rocks also form the area immediately around Stickle Tarn.

From the 'summit', after enjoying the views back from where you have come and towards Blea Rigg, strike out across the grassy moorland towards the summit of High Raise. From here on, until you reach the Langdale Pikes, the terrain is largely featureless and would present difficulties in the mist unless you are confident of your navigational skills. If there is mist or low cloud it would make sense to miss out this section of the walk and make your way round the valley head, above Bright Beck, directly to Pavey Ark where the descent routes are easier to locate in the mist.

Photo w7.2 |

Breccia/coarse tuff with
small lumps (about the size
of a pea), one larger lump
(3cm high) and a similar-sized
patch of yellow 'map' lichen.

Photo w7.3 |

Sedimentary rocks on the
summit of Sergeant Man.

Photo w7.4 |

View north from High Raise.

In clear weather, head towards High Raise. The 'High White Stones' near the summit and others in this area are tuffs and volcaniclastic sandstones. The views from the summit are really excellent (see Photo w7.4), especially when looking away from the Langdale Pikes which are fairly insignificant features from here.

Langdale Pikes

When you have had enough of the super views towards nearby Glaramara and the more distant skyline mountainscape it is time to head towards the Pikes, aiming first towards Thunacar Knott. Instead of heading over the top of Thunacar Knott, as you approach it follow the track which bears left heading towards Pavey Ark. If you take the line of least ascent/descent you will arrive at the Pavey Ark ridge a short distance to the east of the summit, so bear left to head towards the high point. Keep an eye on the rocks as you go, as you will see a very clear and distinct breccia with lumps up to the size of golf balls or larger (see Photo w7.5). As you get closer to the summit, the lumps get bigger and bigger until they become unlike anything else we have seen (Photo w7.6). Geologists are unsure what processes may have caused these weird shapes.

Photo w7.5 | Breccia outcropping near the summit of Pavey Ark. Bedding can be seen in the upper left and breccia in the lower right corner, with lumps up to about 7.5cm across.

Photo w7.6 | Breccia outcropping near the summit of Pavey Ark. The largest lumps are much bigger than those in Photo w7.5, at about 40cm across (taken a few metres from Photo w7.5).

From Pavey Ark head towards the summit of Harrison Stickle, the highest point among the Langdale Pikes at 736m (Sergeant Man is 734m and High Raise tops 762m). The height and pivotal position of Harrison Stickle ensures excellent views and the slabby nature of the summit rocks also provides quite a few good sheltered spots in windy conditions. More breccias are seen around the summit and on the descent.

The direct route down from Harrison Stickle involves a couple of short but steep scrambles over rock outcrops. If you want to avoid this, drop down from the summit to the col between Harrison Stickle and Pavey Ark, then bear left to head down an easier track. Whichever way you drop down from Harrison Stickle, cross the boggy basin aiming for Pike o'Stickle and scramble up to its summit for more fantastic views among the outrops of pyroclastic and sedimentary rocks. The views of the crags forming this southern rim of the Pikes show very clear sedimentary bedding (see Photo w7.7).

Langdale Pikes

Photo w7.9

View of

Dungeon Ghyll.

From the summit of Pike o'Stickle scramble carefully down to return to the track and then head south-east towards Loft Crag. The col at the base of Pike o'Stickle is the site of an ancient stone axe quarry whose products have been found in various places across Britain (see Photo w8.1).

The walk towards Loft Crag takes you past a variety of pyroclastic rocks, mainly varieties of rhyolitic tuffs and breccias erupted as pyroclastic fragments in ash-falls and ash-flows. Some of the eruptions were so hot and gaseous that fragments of rock around the vent were torn away and welded into the fabric of rock laid down by the erupted material, forming welded tuffs (see Photo w7.8).

From Thorn Crag drop down to pick up the path that descends from below Harrison Stickle on the northern side of Dungeon Ghyll. The first five or ten minutes as you descend, passing through the crags rearing up on either side of you and just after, are sensational. If you want to appreciate the views, do stop and look around as you need to keep your eyes on the track as you descend here and a trip would be serious. Dungeon Ghyll drops dramatically off to your right (see Photo w7.9).

The sensational section soon finishes and is replaced with a simply superb walk down to Pike Howe. The track turns off right just before reaching Pike Howe, but it is worth carrying on the few metres to the top of Pike Howe for excellent views. Then return to where the track descends and drop down the steep path back to the car park/bus stop.

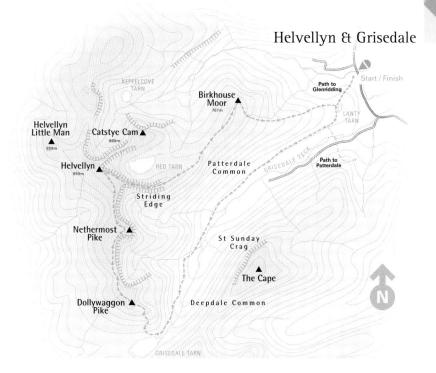

Helvellyn & Grisedale

Walk #8 Helvellyn & Grisedale

START	▶	GLENRIDDING CAR PARK/BUS STOP (387 169)
FINISH	●	CIRCULAR ROUTE
TIME	⏱	6 HOURS+
GRADE		NAVIGATION ● ● ●
		TERRAIN ● ● ● ● ●
		SEVERITY ● ● ● ● ●

Helvellyn is without a doubt one of Lakeland's best loved mountains – and with good reason. Its eastern side presents some of the most dramatic glacial scenery in England and, with Striding Edge, it also offers one of the most exhilarating hillwalking routes for getting to see that scenery.

Helvellyn & Grisedale

This walk follows Striding Edge to the summit of Helvellyn, then heads south to follow the edge of the ridge line with stunning views over a series of corries, arêtes and other glacial scenery carved into tough pyroclastic rocks, finishing with an atmospheric walk down the moraine-strewn glacial valley of Grisedale.

It is quite a long walk, but once the summit ridge is attained the bulk of the hard work is done. A shorter option would be to descend via Swirral Edge and Glenridding (with an optional visit to Catstycam), but the full walk presents an unrivalled view of the sensational glacial scenery of Helvellyn and upper parts of Grisedale.

Helvellyn sits near the middle of one of the great mountain ridges of Lakeland, stretching from Clough Head in the north via Great Dodd, Stybarrow Dodd, Raise, White Side, Helvellyn and Nethermost Pike to Dollywaggon Pike. A southern limb of the same range continues south, splitting at Fairfield into a western limb (Great Rigg and Heron Pike) and an eastern one (Hart Crag, Dove Crag, High Pike and Low Pike).

The eastern side of the range is bounded by a major fault line (one of the biggest in Lakeland), the Coniston Fault.

With a major range and a major fault line, this area is one of the core structural features of Lakeland geology. The northern fells in the range (above Helvellyn Lower Man) are mainly, but not exclusively, lavas of the Lower Borrowdale Volcanic Group. From the south, they are made up of pyroclastic rocks from the Upper BVG (including sedimentary rock deposited between episodes of eruptions).

The Helvellyn range exemplifies the different ways glacial effects can shape the uplands. The western side of the range is rounded and grassy, providing easy walking and grand views of the distant ranges west and north, but it's not exhilarating. There is a restrained atmosphere, such as someone determined to remain on good behaviour and show only their smooth manners.

Cross to the eastern side of the range and the scenery is utterly transformed. Here the rocky excesses are uninhibited and shamelessly show off a savage side to this landscape. This walk explores that rough, rocky aspect of Helvellyn, its neighbouring peaks and the glacial valleys, corries and arêtes that underlie that savagery.

The glaciers which sat in the corries seen from above on this walk were among the last in Lakeland to melt at the end of the ice age (along with those on the eastern flank of the other great north–south ridge, the High Street range), while the corrie

Photo w8.1 | View of sedimentary rock in the Deepdale Formation on Striding Edge
(foreground to centre distance) and the summit area of Helvellyn (top centre distance).
Pyroclastic rocks in the Helvellyn formation form the side walls of the corries on either
side of Striding Edge, best seen in the crags above Nethermost Cove (left).

glaciers in the west of the area were the first to disappear. Thus the eastern flank of
Helvellyn is the location of some of the best corries in Lakeland as there was more
time for the corrie glaciers to do their sculpting of the fell landscape.

Follow the road on the left of the river bridge past a row of shops and houses, bearing left
after about 250m towards Keldas and Lanty's Tarn. It is well worth taking the five minutes
needed to follow the permissive path out to Keldas for the view over Ullswater. Return to
the footpath and continue past Lanty's Tarn, then immediately after the tarn turn right on a
minor track that heads up to a stone wall.

Cross the stone wall, turn left and follow the track up towards the summit ridge of Birkhouse Moor. Some way before the summit ridge, the main track from Glenridding via Mires Beck
comes in from the right. Shortly before meeting it (at about 373 161), you pass an outcrop of

pyroclastic rocks and tuffs, well-cleaved in places and in others with coarse tuffs and breccia.

This is what geologists call the Helvellyn Formation, largely consisting of dacitic tuffs and coarse tuffs erupted some time after the eruptions that created the Scafell caldera, and thus among the later phases of the Lakeland eruptive cycle. The pyroclastic eruptions that laid the Helvellyn formation rocks were highly explosive. However, the material that was erupted didn't travel very far and was largely contained within a 'fault-bounded depression' or caldera of some sort in this area. The rock layers are quite thick (up to 300m) but not widespread. These rocks cover much of the area around this walk, but not on Striding Edge nor, despite their geological name, the summit of Helvellyn.

The navigation here becomes fairly straightforward, being largely a case of following the track. However, watch out for where the track moves away from the stone wall to include a visit to the 'summit' of Birkhouse Moor, then heads back towards the stone wall leading to Striding Edge. The rocks as you approach Striding Edge (and on the Helvellyn summit ridge) are from a different group of rocks: the Deepdale Formation. This is a mix of tuffs and 're-worked tuffs' laid down as volcaniclastic sandstone in a large lake (thus accounting for the bedding which is seen in some of the outcrops) in the depression created at the time of the Helvellyn formation eruptions.

The first two phases of a cycle of caldera creation typically include: first, a violent pyroclastic eruption leading to the collapse of a land mass bounded by faults and second, sedimentary deposits of tuffs from further eruptions and from 're-working' of previously eroded material. These phases are represented in this area by the Helvellyn and Deepdale Formations respectively. There is often a third declining phase in which quieter lava or pyroclastic eruptions occur, but this is not apparent here.

The track leads you towards the narrowing ridge line of Striding Edge. The best route is to stay right on top of the ridge with sensational views down to Red Tarn on the right and Grisedale on the left. However, there is a track below the ridge line on the right which should be followed by those who find the ridge itself too vertiginous and in high winds by all sensible walkers. Though often described as 'knife-edge', the ridge is in fact quite wide (compared with Crib Goch on the Snowdon horseshoe) and should be fairly safe except in high winds. The most difficult part is a short descent at the end of the ridge where it meets the main eastern face of Helvellyn; here it is necessary to turn in and face the rock to climb down a rocky crag. Again there is a by-pass track (this time on the left). It is, however, unpleasantly difficult (not to say dangerous) where it returns to the main track, so it is best to avoid the by-pass track unless you find the direct descent too difficult.

Photo w8.2 | A classic corrie with a glacial tarn and an arête on either flank:
Swirral Edge (left), Red Tarn (centre) and Striding Edge (right).

The Striding Edge ridgeline is one of the best glacial arêtes in Lakeland, a sharp ridge created by glaciers in the corries on either side, cutting back into the intervening rock until only a narrow ridge is left behind (see Photo w8.1).

The deeply engraved corries which run along the whole Helvellyn ridge are all on the eastern side, while on the summit and to the west the slopes are smooth and unbroken. This is because snow collects more easily on the colder eastern sides of the mountains where there is little sunshine and shelter from the prevailing westerly winds. If these conditions allow the snow to persist and turn into glacial ice, glacial features such as corries, U-shaped valleys and arêtes are carved out on the eastern side. Indeed, the flattish summit ridge of Helvellyn may well represent a surviving landform from the pre-glacial surface of gently sloping hills and valleys.

The walk along Striding Edge and the scrabble up to the summit present you with superb views into the corrie below Helvellyn and its glacial tarn, Red Tarn. It is impounded by moraines left behind in its dying stages by the corrie glacier that once existed here.

After the almost level walk along Striding Edge itself, the track becomes suddenly steeper once the ridge is passed and your lungs and legs are rudely forced into full action. Don't rush

Helvellyn & Grisedale

Don't rush this section, but make use of it to stop and turn round as often as you can to appreciate the excellent views of Striding Edge as you ascend stage by stage (see Photo 7.10). Eventually the steep slope flattens out and you are cast onto Helvellyn's flat, small boulder-strewn summit plateau.

The summit shelter is to the right. The views from here are more restricted than those you get from the edge of the plateau overlooking Red Tarn, but the cross shape of the shelter does at least offer some way of getting out of the wind which otherwise whips across the summit plateau unimpeded by anything bigger than a large pebble. (However, there is often little choice about which bit of the shelter you get to use given the number of people thronging around it.)

From the summit shelter, after enjoying the widespread views (and, if the weather permits, taking a short diversion north to overlook the descent via Swirral Edge arête to Catstycam), head south towards Nethermost Pike. The main track bypasses the summits of Nethermost Pike and Dollywaggon Pike, presumably to avoid the small amount of extra ascent involved in visting them. These are really not individual mountains, being little more than slightly higher points along the ridge, so many people miss them out.

In fact, these high points mark the places on the ridge where it meets the top of the arêtes that have developed in between the corries south of Striding Edge. The main track therefore derives you of an utterly marvellous 2km walk above one of the most striking natural geological/glacial features in Lakeland, the run of corries between Red Tarn and Grisedale Tarn. So, unless forced to by mist, tiredness or high winds, don't miss the electrifying views down into the corries of Nethermost Cove and Ruthwaite Cove (see Photos 7.2 and 7.3).

Follow the edge of the ridge line all the way until you start to descend from Dollywaggon Pike to Grisedale Tarn, where the zig-zag track should be followed to avoid adding to the erosion of these slopes.

On the walk southwards along the ridge edge, at the slight dip beyond Nethermost Pike, you pass from the Deepdale Formation rocks back to the Helvellyn Formation (as seen earlier on Birkhouse Moor) and then back into the Deepdale Formation again briefly around the summit area of Dollywaggon Pike. Descending towards Grisedale Tarn, you pass intrusive andesite or lava with some flow-banding and breccia to be seen (Photo w8.3). When you get down into Grisedale, you again meet pyroclastic rocks and follow them all the way back to the start point. Some more andesite is passed right at the end of the valley. Examples of welded tuffs (photo w8.4) can also be seen.

Photo w8.3
Brecciated lava,
head of Grisedale.

Photo w8.4
Welded tuff, head
of Grisedale.

The route down through Grisedale is a superb walk, passing plenty of interesting rock outcrops and great clusters of moraines, waterfalls and old mining works. The track takes you down to Ruthwaite cottage, where shortly afterwards you have a choice of route.

Crossing on the footbridge (362 143) to the south-east (on the right as you descend) side of the valley, along a farm track on the valley floor, offers the easier walk. Sticking with the north-western side (on the left) is more interesting, however, and offers generally better views as it uses a narrow track some way up the valley side (it also crosses a footbridge but over a side stream, Nethermost Beck).

The advantage of the right-hand track, apart from being easier walking, is that it takes you right through a complex of moraines at around 365 146. The advantage of the left-hand track, apart from the variety of rocks and views, is that it is possible to get a good look at the

Helvellyn & Grisedale

Photo w8.5

Waterfall at a
rock step on
a geological
boundary
between two
rock types,
upper Grisedale.

waterfall (about 363 145) by leaving the track and heading towards the waterfall with great care. Also, the left-hand track leads directly to Lanty's Tarn and the end point at Glenridding. The right-hand track requires you to turn left at about 383 157 up a track to rejoin the left track (at about 3815 1595) before returning to the end point.

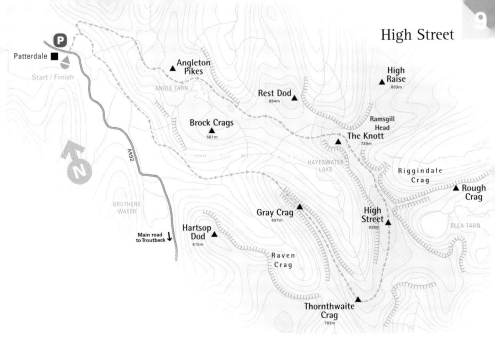

Patterdale ■
Start / Finish

ANGLE TARN

Angleton
Pikes ▲

High
Raise ▲
803m

Rest Dod ▲
694m

Brock Crags ▲
561m

Ramsgill
Head
The Knott ▲
739m

HAYESWATER
LAKE

Riggindale
Crag

▲ Rough
Crag

BROTHERS
WATER

Gray Crag ▲
697m

High
Street ▲
828m

BLEA TARN

Hartsop
Dod ▲
615m

Main road
to Troutbeck ↓

Raven
Crag

Thornthwaite
Crag ▲
783m

Walk #9 High Street

START	▶	JUST SOUTH OF PATTERDALE AT SIDE ROAD JUNCTION (398 158)
FINISH	●	CIRCULAR ROUTE
TIME	◔	6 HOURS+
GRADE	⊙	NAVIGATION ● ● ● ●
	☁	TERRAIN ● ● ●
	✕	SEVERITY ● ● ● ● ●

Going for a walk along the High Street takes on a wholly different meaning when it involves a large hill of that name, a hill which is the highest point on the majestic ridge that defines the structure of the eastern fells of Lakeland.

The ridge runs for about 20km (12 miles) in a virtually straight line, trending almost directly north–south. The route of an ancient Roman road (which does how-ever bypass the high point) runs along the ridge for a good part of its length. The

High Street

walk from one end of the ridge to the other is one of Lakeland's great fellwalking challenges, but it is not suitable for our purposes as it leaves little time for studying the rocks and views. The walk described here visits the highest point on the ridge and also takes you past some dramatic glacial scenery. It is still a long and demanding walk, but a rewarding one.

The ridge itself is broad and wide, but the views are truly excellent. However, the shape of the ridge also means that, unless you are experienced and confident in bad weather navigation, this walk is best avoided in misty conditions. It would be all too easy to get lost on the descent and end up on difficult ground or to drop down into the wrong valley. In addition, the main ridge is not a simple linear feature but quite complex with lots of side ridges and slight twists of direction; its very large scale could be confusing. This means that even in good weather, careful attention to the map and ground is essential.

Today's flat-to-gently rounded summit plateaux may be remnants of the pre-glacial surface. This effect is seen most clearly from the summit area of High Street itself, looking south and east to neighbouring fells.

The corrie glaciers in this part of Lakeland were among the last to melt at the end of the ice age. This is because it is slightly colder farther away from the temperate influence of the sea that aided earlier melting of glaciers in the west. These valley and corrie glaciers have cut deep-sided valleys into the mountain range with steep valley heads and broad separating ridges. Some classic corrie features exist on the eastern side of the ridge below the summit of High Street, Blea Water.

It is interesting to note how the glacier that carved Blea Water left behind conditions suitable for the growth of a lake, whereas the adjacent glacier in Riggindale did not result in a lake. This underlines the point that corrie lakes only develop under suitable circumstances dependent upon the interplay of rock type, rock boundaries and the presence or absence of faults. Riggindale follows an area of andesitic lava and a fault, both running along the valley itself, with rhyolitic tuffs forming much of the higher land on either side of the valley. Blea Water is held within a small area of one type of rock (andesitic and dacitic tuffs), but is surrounded by different types of andesitic lavas and some rhyolitic tuffs. All the rocks on the eastern and northern flanks of the High Street range are part of the Lower Borrowdale Volcanic Group.

Photo w9.1 | Approaching Angle Tarn. The lake is enclosed in a dip carved by glaciers exploiting weaknesses in the rock outcrops seen around the lake.

Despite the appearance of simple, rounded topography, the geology of the area is highly complex. We have been left with an interplay of rock type and faults which have thrust different types of rock up or down alongside one another during the convulsions of the volcanic eruptions. The ridge above Hayeswater, on the descent of the described route, looks little different from other summits and ridges in the area, but in fact marks the boundary of one of the great calderas created by pyroclastic eruptions in the Upper Borrowdale Volcanic Group.

Follow the side road over the bridge to the other side of the valley. Follow the road when it bends left, heading up to a farm gate across the road, where on the right there is a gate giving access to the footpath up to Boredale Hause. The track rises gently, slanting up the hillside, passing tuffs, welded tuffs and volcaniclastic sandstone, eventually bringing you up to the complex col and its multi-choice junction of tracks.

The walk follows the track on the right, heading into a slight valley heading up towards Rake Crag and past outcrops of well-cleaved volcaniclastic sandstone and then andesitic lavas. Higher up there is a choice of an upper and a lower path; they rejoin near Angle Tarn which comes into view as you make your way around the lower flanks of Angletarn Pikes (see

High Street

Photo w9.1). Columnar jointing (formed by the slow cooling of intrusive magma) can be seen on the other side of Angle Tarn. This is a popular spot to stop and have a break before the final parts of the ascent.

From Angle Tarn the track heads towards Satura Crags. This section of the walk offers unrivalled views of Hayeswater and its deep glacial valley (see Photo w9.2). Here it is important to avoid being drawn up towards Rest Dodd. Ensure that you follow the right-hand track, which goes around the south-western flanks of Rest Dodd towards the main ridge and The Knott and then to the grandly named Riggindale Straits.

These straits are not very narrow (in mountain ridge terms) but there is an important junction at the straits with the main track (the old Roman road) coming in from the north and south. As soon as you meet this junction, leave the main track and head to the eastern edge of the ridgeline for superb views of the glacial corries and arêtes below High Street. Make your way up along the edge towards the summit (except in very strong westerly winds when this could be dangerous). The walking is a bit harder than if you stick with the track, but the views more than repay the small amount of extra effort. This is especially true as the Long Stile arête gets nearer. As you pass above the arête, note the tiny tarn suspended in a slight dip as well as the classically circular-shaped glacial corrie and its impounded glacial lake, Blea Water (see Photos 7.5 and 7.6).

Photo w9.2 | High Street (upper left), Hayeswater (centre) and Gray Crag (upper right): a classic glacial valley with considerable amounts of vegetation-covered scree covering the slopes on either side. Hayeswater is impounded behind moraines.

Photo w9.3
Looking north from
the summit of
Thornthwaite Crag
towards Ill Bell (left)
and Windermere
(centre distance).

The lake is pretty deep, as much as 63m at its deepest point. Even that understates the depth of the glacier that it once held, for the distance between the deepest point of the lake and the highest point on the moraines that sit in front of it is nearly 100m. The moraine material immediately in front of the lake is only superficial. It rests on top of a rock bar which impounds the lake, underlining the point made by geologists about how corrie glaciers can dig down into the surrounding rock to create deep basins. This means that the ice had to flow uphill to some degree to escape from the rock basin before it could flow downhill. The 'lateral' moraines formed by the corrie glacier along its sides can be clearly seen. The glacier in this corrie would have been one of the last to melt at the very end of the ice age, sheltered and cold in this eastern stretch of the area.

Blea Water is one of a cluster of corrie and valley glacier sites around High Street which housed glaciers that were active in the final stages of the ice age. The glacier from the Blea Water corrie merged with another glacier in Small Water corrie, and there was another east-flowing glacier in Riggindale (passed on the way to this point). There were north-flowing valley glaciers in the Hayeswater and Threshthwaite valleys to the west as well as others.

Ice has certainly carved its way deeply into these hills, creating the superb scenery all around. On the other hand, the ice has largely left the summit surface untouched and it is probably a remnant of the pre-glacial surface. At the height of the ice age it would have been buried under the ice sheet that covered the area. At the end of ice age when corrie glaciers were actively carving the landscape, only periglacial freeze-thaw actions would have been effective in shaping the landscape. This has left us with the broken boulder fields and subdued summit tors we see today.

High Street

Photo w9.4
Showers over the
Helvellyn range seen
from Grey Crag.

The flat summit of High Street is near the stone wall that runs along the ridge. The open aspect makes this a pretty windy place; however, the stone wall can often offer shelter depending on the wind direction. From the summit, head towards the slight depression and rise to Thornthwaite Crag (see Photo w9.3). The best views are had by leaving the track and going to the edge to peer down into Hayeswater where the moraines left by the glacier are easily visible on either side of the lake.

Return to the track and on to Thornthwaite Crag and its upright cairn. From here, head just east of north along the delightful ridge (an arête between two glacial valleys) towards Grey Crag, with excellent views of the moraines around Hayeswater. All too soon this lofty ridge walk (see Photo 7.12) comes to an end, and a steep descent follows. As you near the bottom of the ridge look out for a small footbridge near some water treatment works (at 422 130).

The direct descent to the footbridge is too steep and it must be approached either from above right or below left. Cross the bridge and pass the water works, picking up a track descending into the valley. Bear right almost immediately along an initially boggy track that rises gently to above the intake fence/stone wall. Note the structures lower down in the valley on the neck of land where the two streams merge. No doubt it was used to guide water to work machinery used by the mining operations that can be seen around the area.

After a few hundred metres the track starts to descend down a stony track, for a while steeply. Farther on a way has to be found across a stream. If the stream is too fast to walk across, scramble across boulders next to the fence on the left down to a footbridge. The track eventually leads you back to the minor road at 400 160 and thus back to the start point.

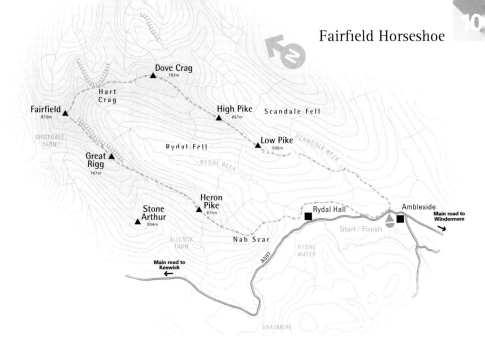

Walk #10 Fairfield Horseshoe

START	▶	AMBLESIDE TOURIST INFORMATION OFFICE (377 045)
FINISH	●	CIRCULAR ROUTE
TIME	☕	5 HOURS+
GRADE	⌂	NAVIGATION ● ● ●
	⛰	TERRAIN ● ● ●
	⊗	SEVERITY ● ● ●

The Fairfield horseshoe is rightly a very popular walk, linking two long slender ridges with a superb mountain summit area. The ridges are narrow enough to ensure excellent views in all directions, but broad enough (except at one or two steep points where height is gained or lost) to be a relatively easy mountain walk. (Note that this statement does not allow for potentially awful conditions in gales on exposed ridge tops.)

The summit area of Fairfield itself is rather flat, but by going out to the edges of the summit plateau and peering over the crags you are presented with a stunning

Fairfield Horseshoe

array of mountain scenery. Grisedale Tarn, Dollywaggon Pike, Striding Edge, Grisedale and St Sunday Crag command the attention from Fairfield's northern limits, a miscellany of tough volcanic rock carved into intensely dramatic glacial features. There are plenty of moraines to be seen, especially in Deepdale and Scandale.

There are some interesting breccias and other volcanic rocks on and around the summit and other geological features, including sedimentary volcaniclastic sandstone, are revealed on the descent part of the walk.

The tough decision facing anyone who wants to walk the horseshoe is to decide which way to tackle it: clockwise or anti-clockwise. Both directions (and other options too if you bring the Red Screes ridge into the range of possibilities) have considerable advantages and no disadvantages. Going clockwise means a good walk-in to loosen the muscles before starting to climb, a short steep ascent to gain the ridge followed by an easy ramble to the summit and a gloriously long and gentle descent towards Ambleside and Windermere. Going anti-clockwise means a long gentle ascent and a lofty descent looking towards the fells around Grasmere. I have chosen to go clockwise, but the choice is arbitrary.

The fairly frequent bus service between Ambleside and Keswick makes it a practical proposition to shorten the walk by using the bus to cover the leg between Ambleside and Rydal.

For the greater part of the route the navigation is exceptionally easy: simply follow the ridge line. However, the summit area (which encompasses Fairfield, Hart Crag and Dove Crag) is rather broad and fairly flat and could present difficulties, especially in the mist, if insufficient attention is paid to map and compass. The track is well worn, but there are many different tracks heading off the summit at various places and it would be easy to go in the wrong direction.

Follow the road towards Rydal as far as 372 052 where a public footpath leads you towards Rydal Hall. Pass through the grounds, noting the hummocky glacial terrain on your left once you get away from the road. Pass through the cluster of buildings around Rydal Hall (with café) to join minor road at about 3647 0632. Turn right uphill along the road, which soon becomes a public footpath, to beyond Rydal Mount. Bear left towards the steep track up Nab Scar. You are soon landed on the ridge. Nearly all the rocks you come across are tuffs and

Photo w10.1 | Looking along the ridge towards Great Rigg and Fairfield; Rydal Beck (lower right) and the return route along the Dove Crag/High Pike ridge (upper right skyline).

other pyroclastic rocks (although there are a few small outcrops of intruded andesite). You can see some cleaved tuffs as you reach the start of the ridge and as you climb gently up towards the high-points on the way.

Make sure you take time to enjoy the unfolding views all around, including the view back towards Ambleside and Windermere and on either side of your route. Note the many moraines left by retreating glaciers in the valley on the right, drained by Rydal Beck.

The route takes you over Heron Pike and Great Rigg. There is nearly always a good view forwards as you climb progressively over the outer peaks (see Photo w10.1). The views can be quite dramatic even in cloudy weather (Photo w10.2). Carry on to the summit of Fairfield, a wide and rather flat field of small boulders.

As mentioned earlier, it is recommended that you take time to stroll along the edge of the summit plateau overlooking Deepdale, St Sunday Crag, Grisedale and Dollywaggon Pike for the interplay of mountain and deep glacial valley. The row of glacial corries between Dollywaggon Pike and Helvellyn is best seen from St Sunday Crag rather than from here. The profile of the arêtes separating the corries, with Striding Edge forming the skyline, is

Fairfield Horseshoe

Photo w10.2 | Looking back to Nab Scarr as cloud slowly lifts to reveal views of the walk so far.

especially noteworthy. This is some of the most intensely glaciated upland scenery in the Lake District (see Photo w10.3).

A visit to St Sunday Crag is worthwhile for the view of the corries and towards Ullswater, but the descents and re-ascents involved are quite demanding and will ensure tired legs on the long return from Fairfield to Ambleside.

To return to Ambleside from Fairfield, ensure that you pick the right direction for the descent. First, head south-east for about 150m, then swing east for about 500m. The track then starts to swing to a south-easterly direction to Dove Crag, where it then trends south.

Around where the change of direction from south-east to east occurs, look out for the rock outcrops on the left where breccias can clearly be seen (see Photo w10.4).

After Dove Crag, the ridge takes on a form which will accompany you for the next 2.5km: a narrowing, gently descending finger of rock extending deep towards Ambleside, accompanied on either side with narrow steep glacial valleys and another ridge line beyond that.

There is a steep section of descent a couple of hundred metres after the inconspicuous summit of High Pike (also known as Scandale Fell), with clear bedding of the rocks especially on the left.

Photo w10.3 | St Sunday Crag (left), an arête created by adjacent glaciers in Deepdale (right) and Grisedale. Hummocky moraines can be seen in the lower centre of the picture in Deepdale.

A bit lower there is a change in the formation or rock type (the Esk Pike formation higher up and the Lingcombe Tarns formation lower down). There is another change in rock formation to more sedimentary rocks in the Seathwaite Fells formation lower down. Farther along, when you get to the next flatter (or undulating) section, look over to the left to the

Photo w10.4 |
Breccia seen a short
distance to the left
of the track.

Fairfield Horseshoe

Photo w10.5 | Scandale: note moraines crossing the valley and also the layers of rock exposed in the slopes (upper right).

other side of the valley to see layers of crags (see Photo w10.5). These match up with the next set of crags you descend through a bit farther on where the track swings left below the corresponding crags. The track then swings right (alternatively, stick to the ridge line and then scramble down crags farther on). Eventually you are brought out to Low Sweden bridge and a minor road back to Ambleside.

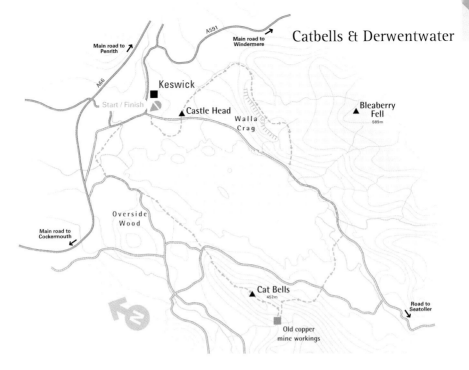

Catbells & Derwentwater

Keswick

Main road to Penrith

Main road to Windermere

A591

A66

Start / Finish

Castle Head

Walla Crag

Bleaberry Fell
589m

Overside Wood

Main road to Cockermouth

Cat Bells
452m

Old copper mine workings

Road to Seatoller

Walk #11 Catbells & Derwentwater

START	▶	KESWICK BUS STATION (263 235)
FINISH	●	CIRCULAR ROUTE
TIME	🕐	5 HOURS+
GRADE	ⓝ	NAVIGATION ● ●
	⛰	TERRAIN ● ●
	ⓢ	SEVERITY ●

If the domestic moggy is but a manageably small version of the wild jaguar, then Catbells is a petite version of Lakeland's more muscular northern fells (or at least of those formed of sedimentary rocks). This miniature mountain mimics the slopes and shape of its more substantial near neighbours, even sporting its own shapely summit cone. The delightfully easy ascent of Catbells is hardly even a mountain climb

Catbells & Derwentwater

(although the short climb is fairly steep and involves a small amount of scrambling at one point). An ascent of Catbells is hardly sufficient to fill a half day, let alone a full one. However, it is definitely an ideal way of gaining some height on a day when the taller fells have their heads stuck in the mist.

Combine it with a tour round the edges of Derwentwater, an optional walk over the top of the lava crags of Walla Crag and then a short traverse atop the exposed plug of an ancient volcano, Castle Head, and you have a very fine day's outing on and around a variety of classic Lakeland rock types and glacial features.

The walk takes you over sedimentary rocks of the Skiddaw Group and also the lavas of the Lower Borrowdale Volcanic Group, as well as offering tempting views towards the pyroclastic terrain of central Lakeland. The day's walking is however dominated by Derwentwater and the views of the mountain landscape which frame it to such perfection (see Photo w11.1). Derwentwater is a fairly wide lake spreading out across the area between the sedimentary Skiddaw Group hills of Catbells and Maiden Moor on the western side, and lavas of the Lower BVG on the eastern side. The pyroclastic rocks of the Upper BVG are temptingly glimpsed forming the skyline beyond the much admired Jaws of Borrowdale. Here, the valley narrows considerably compared to the wider valley farther north. The constriction marks where the tough Lower BVG rocks cross the valley.

From the bus station head initially north-west to the A591 road, turning left onto it and then cross the bridge over the river Greta. Immediately after the bridge, turn left onto a public footpath. Follow the path to a minor road and a bridge across the river Derwent and then into Portinscale, taking the minor road south out of the village for about 1.2km. Bear left on a public footpath through the woods (still heading roughly south and thus avoiding the footpath which heads off east near here). Follow the path through the wood to the base of Catbells and then pick up one of the tracks heading up the north ridge from above the cattle grid marked on the OS 1:25,000 map at 247 212.

A stiff climb of some 200m takes you to the first step in the slope. Shortly before the climb steepens again you pass a narrow trench (an old quarry working) passing across the track from one side of the hill to the other. If you take a few minutes to follow the cutting in both directions you will see that the workings extend downhill (see Photos w11.2 and w11.3).

Photo w11.1 | Derwentwater at dawn (looking south from near Keswick) and Catbells (extreme right).

These are the remains of an early copper mine which was first worked in the 16th century by miners from Germany.

On the final short stiff ascent to the summit the track takes you close to outcrops of the well-cleaved sedimentary rocks of the Buttermere formation, part of the Skiddaw Group (see Photo w11.4). These cleaved rocks are siltstones and sandstones that have been heavily folded

Photo w11.2 | Scars left by early quarrying operations on the lower ridge leading up to Catbells (looking south-east).

Photo w11.3 | Quarrying operations scars on the ridge leading to Catbells (looking north-west).

Photo w11.4
The short scramble
through sedimentary
rocks near the
summit of Catbells.

and subject to shearing and disruption during mountain-building episodes. From the summit there are superb views west to Causey Pike (see Photos w11.5 and 1.8, which show the view from Hause Gate after descending from the summit). There are also excellent views of the range of north-western fells and around to the south-east to the Jaws of Borrowdale and the northern end of the ridge that runs continuously south to High Raise and the Langdale Pikes (see Photo 2.1). Similarly the views towards the Skiddaw massif are enthralling.

The view to the south and east is to volcanic terrain par excellence, knobbly and irregular. Causey Pike and the north-western fells, on the other hand, exemplify the smoother, more regular terrain carved into the softer sedimentary hills with their narrow pointed, level ridges.

From the summit head down south to the col, Hause Gate, keeping an eye out for a tiny mine entrance on the slope leading down from Maiden Moor. It is just above the col and to its right. There is a small track leading towards it and there is also a small waste tip there. Don't

Photo w11.5 | Classic ridge in Skiddaw Group rocks: Causey Pike (centre), Grasmoor (centre left distance), Grisedale Pike (right distance) and Newlands valley (foreground).

enter the mine as it may be unsafe, but you can peer in and wonder at the bravery of men who went to work in the pits of earth (see Photo 8.6). If you pick your way through the waste tip you'll come across small bits of rock containing coloured quartz and metallic minerals. This is what the miners were seeking: copper-bearing veins of quartz. Looking down the hillside

Photo w11.6 | Viewing the scene from the summit of Catbells.

Catbells & Derwentwater

opposite (below Catbells) you can see several more waste tips in the immediate area and even more of them lower down the hillside (see Photo w11.7).

The mines clustered around here began operations in the 16th century with German miners, but their heyday was in the 19th century when there was a major mining operation based on this set of sites and in the Newland valley below you. Take care not to enter any mining levels of shafts and watch your step if you study the rocks on the waste tips.

Start to return to the col and be sure to look upwards at the crags on the slopes leading to the summit of Catbells. The valley side has been cut across the dip of these rocks so that they present a roughly cut face when they poke through the scree and grass as an outcrop. When you start to descend on the other side towards Derwentwater you will see that the slope there parallels the bedding, so that the rocks, where they outcrop, present a smooth face outwards (see Photos 1.9 and 1.10). This phenomenon is also seen on many other fells in the Skiddaw Slates area, especially among the north-western fells (see also Walk 1).

Follow the path down to a minor road at about 2505 1850 and follow the road south for about 350m, where you turn left onto the public footpath that heads north-east towards the southern end of Derwentwater. The ground here is very wet as this was once part of the lake and is only slowly being transformed into land (thanks to the National Trust the terrain is not a problem to cross as they have provided a wooden walkway over the wettest sections).

Keep an eye out to the south as you cross the flat valley floor for the excellent views of the Jaws of Borrowdale, the boundary between the sedimentary and volcanic rocks (see Photo 2.9). In the other direction you get a superb view north across Derwentwater to the Skiddaw massif (see Photo 6.5). The path brings you out onto the B5289 at about 2625 1870. As you approach the road you get excellent views of the crags ahead, carved into the lavas emitted by the first volcanic eruptions in the area. A major fault runs along the eastern shore of Derwentwater, marking the point where the sedimentary Buttermere Fells formation (part of the Skiddaw Group) butt up against the volcanic rocks.

A diversion from here up to Lodore Falls and some distance up the Watendlath valley (to about 269 181) is highly recommended to appreciate the rocks and the superb woodland and gorge scenery. This is an example of a hanging valley where a small glacier, fed with ice from Blea Tarn higher up on Watendlath Fell, carved out a small valley which joined the much bigger Borrowdale glacier near its top. When the glaciers melted, the floor of the Watendlath valley was left suspended or hanging some way up the main valley side, resulting in the waterfall and gorge that today grace the scenery.

Photo w11.7 | Mineral mine entrance near Hause Gate
(Causey Pike and Grasmoor in the background).

Return to the main road. Follow the road, pavement-less in places, until past the hotel. Beyond the hotel, cross a small stream on the road bridge, then pick up the permissive track on the right side of the road behind a stone wall. Don't forget to look up at the crags of weathered and eroded lava and screes as you follow the track until you are required to cross back to the other side of the road at about 2675 1950. Follow the short stretch of pavement (or shoreside) until another permissive path is encountered (on the left of the road) at about 2677 1975. (Paths such as these had to be constructed when considerable volumes of motor traffic began to invade the roads of Lakeland, making it impossibly unsafe and absurdly unpleasant to walk on them as had been the tradition. This could only happen where the National Trust owned the land, so elsewhere many roads and routes are now effectively out of bounds for walkers). Follow the track, making sure you turn round for views over to Catbells and the fells to its south, Maiden Moor and High Spy. Here you can clearly see the different terrain in the sedimentary rocks of Catbells and Maiden Moor and the volcanic rocks of High Spy.

Follow the permissive path, merging into the shoreside, up to about 2695 2080 where you are forced to rise up to the road for a few metres only. From here you can either carry on alongside the lake or take the slightly higher and longer route over Walla Crag. The routes re-converge at Castle Head. For the lakeside route, simply follow the shoreline track up to the public conveniences (with ridiculously early closing times) at about 2640 2267. Then follow footpaths to the B5289, cross the road and ascend Castle Head.

For Walla Crag, cross the road and go through the gap in the wall opposite, then pick up the track which initially heads north and swings right to rise up into Cat Gill. The track eventually leads you to a footbridge across the gill and another steep ascent to the top of the crags. Note the waterfall on the right where the gill bends sharp left after you come out of the woods.

Where the slope eases off, there is a stile on the left which allows you views from the top of the crags if desired. Return to the track and continue alongside the fence across a grassy area, with a waterfall to the right. Cross the next stile on the left and follow the track across outcrops of lava to the 'summit' and excellent views, especially of the lava steps in the fells to the east and south-east (see Photos 2.10 and 6.6).

Follow the track which picks a way along the edge of the crags and eventually back to the fence. Follow the public footpath north-east towards Brockle Beck and then follow the minor road to the footpath on the left at 2930 2225. Follow the path to 2720 2292, where you turn left and head for Castle Head (see Photos 1.2 and 2.3). From Castle Head follow the public footpaths (taking care when crossing the main road) to the lake and follow lakeside paths towards Keswick. It is necessary to move away from the lake edge as you approach the substantial caravan sites near Keswick, but the minor road you are led onto leads back to the bus station and town centre.

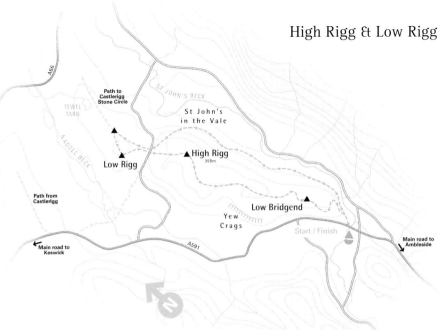

Walk #12 High Rigg & Low Rigg

START ▶ A591 JUST NORTH OF BRIDGE END FARM, PUBLIC FOOTPATH SIGN
(3153 1955). NEARBY BUS STOP ON A591 AND CAR PARK ON B5322.

FINISH ⬤ CIRCULAR ROUTE

TIME ◗ 3 HOURS+

GRADE ◐ NAVIGATION ● ●

◐ TERRAIN ● ●

◐ SEVERITY ● ●

This walk traverses two low hills just to the north of Thirlmere reservoir, just about half-way between the two towns of Ambleside and Keswick. It is a fairly easy walk that can be completed in half a day with minimal amounts of ascent as High Rigg, despite its name, is a pretty low hill reaching just 303m at its summit. As the starting

High Rigg & Low Rigg

point is at 167m, the walk involves little ascent (although the undulating summit plateau does add to the total number of metres that needs to climbed). The low height of the summit means the walk would be suitable when low cloud obscures the higher summits or when a short walk is desired.

An optional extension to nearby Castlerigg stone circle adds another 2 hours plus and is mainly over agricultural land with some resort to minor roads.

High Rigg is largely made up of lava flows and intrusions of basaltic andesite with thin interseams of pyroclastic and some sedimentary rock. This combination of differing types of eruption and differing rock types produces a classic example of 'step' or 'trap' landscape with crags of lava separated by tuffs. This feature is particularly well seen from the minor road (B5332) through St John's in the Vale, especially when heading south (see Photo 2.4), but is also displayed on this walk from a number of different angles. Perhaps the most dramatic view of the lava layers and the stepped slope, however, is seen from below on the return leg of the walk (see Photo w12.4).

The 'up and down' nature of the terrain on the summit plateau gives plenty of opportunity to study the rocks and to appreciate how the initially apparently very random placing of the ups and downs is in fact part of a pattern. This pattern results, like the step topography, from the interplay of the angle of the lava sheets and surface erosion.

High Rigg is a fine viewpoint, surrounded as it is by other ranges. It includes very fine views to the Skiddaw massif in the distance, especially Blencathra. To appreciate the full majesty of the views it is necessary to visit on a day when the bigger hills are also in the clear. On cloudy days this aspect of the walk is sadly missing.

Low Rigg is much smaller and lower than High Rigg and appears to be a simple northern extension of the higher land. However, it is geologically very different to High Rigg as it is made up almost entirely of granite. This is an intrusive rock with rhyolitic geochemical composition which remained deep below the surface, hardly rising at all towards the surface. Instead it cooled very slowly way down in the Earth. The slow cooling means that the rock developed large 'grains'. The magma was probably intruded between 450 and 400 million years ago. Now, a few hundred million years later, it has been exposed at the surface by erosion of overlying layers of rock (see Chapter 4).

Photo w12.1 │ Layers of lava seen on the hillside below Clough Head.

The start and finish point is on the main Ambleside to Keswick road (A591) near bus stops and a car park (which is however accessed from the B5332 a couple of hundred metres north of its junction with the main road). Take the footpath just north of the bridge on the A591, following it for some 50 metres before turning left (north) uphill by some oak trees.

Climb steeply until the first crags you meet next to the path, near where it begins to level out at about 3167 2007. Turning back to look south reveals fine views, especially of Castle Rock (or Triermain), which despite looking like a volcanic plug is an outcrop of rhyolitic tuffs largely surrounded by basaltic and andesitic lavas and intrusions. Lava beds can be seen on the main hillside behind and to the north of Castle Rock (see Photo w12.1).

Looking south-west is the low eminence of Great How sticking out on its own in the middle of the valley. It is made of pyroclastic rocks from material thrown out during the great eruptions that created the Scafell caldera. To the right of Great How, another plug-like eminence is created by The Benn, Smaithwaite Crag and Raven Crag, all part of the Birker Fell formation of lava and intrusive basalts and basaltic-andesites, as are most of the bigger fells to its west (Bleaberry Fell and High Seat). There is also an outcrop of pyroclastic tuffs on High Seat around Shoulthwaite Gill. Again the beds of lava can be seen on Bleaberry Fell.

Turn back to the north and continue towards the summit of High Rigg. Views of the summit plateau start to unfold from this point onwards, revealing a highly hummocky terrain which rises and falls on its slow way towards the summit at the northern end of the plateau, to which you now head. Whichever route you follow requires some degree of walking up and down the crags and hummocks. This topography, with its plentiful 'scarp' slopes and tiny tarns, is the result of glacial erosion exploiting both the varying resistance to erosion of the lavas and interleafed tuffs, and also the presence of many faults.

Trap topography only develops if the angle at which the lava sheets dip away from the horizontal is fairly gentle, allowing erosion to get to work creating the clear steps and rises. Thus trap topography is not guaranteed everywhere that lava sheets are found (as seen earlier looking towards Clough Head, for example).

Most of the rocks you see initially are very heavily encrusted with lichen which has weathered the rocks to grey or pinkish colours, but occasionally you come across the odd rock which has been cracked open for some reason, and you can see the unweathered rock often a dark grey/blue colour.

At about 313 205, near a cairn at a dip between two higher points, head to the right (east) to look over Long Band. This is one of the main lava crags on High Rigg, which will be seen from below later in the walk on the return leg (see Photo w12.2).

Farther on the track bears to the left of a stone wall met in one of the intervening dips. However, if you want to follow the highest points along the route, it is better to keep to the right of the wall and head for the high ground (Photo w12.3). At the next significant dip, and subsequent dips, take time to look back at the rocks in the scarp slopes (i.e. the north-facing slopes).

Photo w12.2 |
Long Band, one
of the lava flows
seen from along
the edge of the
summit plateau.

Photo w12.3 | Terrain produced by the trap topography of lava flows.

Head towards the summit, but just before reaching it bear off slightly to the outcrops to the left to find a good place to sit down and look back at the terrain you've just walked through. Initially it just looks like a messed up mass of hummocks with little or no pattern, but a slightly closer look reveals that there is indeed a pattern. This is best seen when looking back in the medium distance to the scarp slope of the last big dip you crossed. Because the rocks are exposed in this slope, it can easily be traced crossing the summit diagonally (closer to you on the left and further away on the right, Photo 2.8). The grassy hummocks closer to you can also be seen then to fit this diagonal pattern of scarp slopes crossing the summit plateau. The rocks around you here are lavas, with 'vesicular' holes caused by escaping gas shortly after eruption.

Continue to the summit and then start to descend to the road between High Rigg and Low Rigg, aiming to meet it at about 3055 2340 before picking up the footpath to Low Rigg at about 3065 2350. Unless you want to study the granite rocks (Photo 4.9) of Low Rigg in great detail, there is no point in trying to head directly to the summit as a stone wall intervenes. Instead it is best to follow the footpath to a stile in the stonewall at about 304 239. Bear left to the summit where there are plenty of granite crags to be studied. Return to the stile and visit the lower northern 'summit' for superb views towards Skiddaw, Blencathra and also to Bleaberry Fell. It is possible to continue from here using footpaths and minor roads to reach

High Rigg & Low Rigg

Photo w12.4 | The impressive trap topography is seen from below on the return leg of the walk.

Castlerigg stone circle (at 2915 2365) and then return via more footpaths to the road crossing between High Rigg and Low Rigg (see Photo 8.2).

Return to the road and follow it for a very short distance east to the bridleway heading right. Note the lead and iron pyrites quarries on the lower parts of the opposite hillside, Hill Top Quarry to the north, and Bramcrag Quarry and Sandbed Ghyll Quarry to the south. Lava beds are seen above the quarries towards the summit plateau of Clough Head.

As the path rounds the northern head of High Rigg, entering St John's in the Vale, a view is suddenly opened up of the lava steps or traps on the eastern flank of High Rigg on the right. This is quite a dramatic moment and displays the layering of the lava sheets very well indeed (see Photo w12.4).

After a while the path drops down towards the valley floor, getting quite close to the river at one point, with the crags on the right coming close above you. Around here you walk across a lot of lumps of the dark basaltic and basaltic-andesite rocks.

Note the small stone bridge of traditional design across the river near here. Keep on the same track until it passes the farm house (and café) and then starts to rise through a narrow gorge between High Rigg on the right and Castle Rock on the right, before descending and then bearing right to take you back to the starting point.

Walk #13 Lingmoor Fell

START	▶	BUS STOP AT ELTERWATER WITH NEARBY CAR PARKS (328 047)
FINISH	●	CIRCULAR ROUTE
TIME	☕	5 HOURS+
GRADE	⌢	NAVIGATION ● ● ●
	⌢	TERRAIN ● ● ●
	✕	SEVERITY ● ●

This is one of those great medium-level fell walks which can be found in Lakeland, tucked in between the more substantial peaks. However, the feel of the walk along the summit ridge of Lingmoor Fell and beyond to Side Pike is of real mountain terrain. The views are stunning. In my opinion, the vista from the summit ridge of Lingmoor Fell towards the head of Langdale, with Pike o'Blisco, Crinkle Crags, Bowfell, Rosset Pike,

Lingmoor Fell

the Langdale Pikes and Blea Rigg spread around the horizon like a mountainscape amphitheatre, is one of the best views in the Lakeland fells.

The walk also scores high in terms of geological interest and there are plenty of absolutely fascinating rocks, as well as evidence of slate quarrying old and new, to be seen all along this modest mountain. The walk provides opportunities to look through an old quarry waste tip for small chunks of rock with some of the most beautiful rock patterns you will find anywhere, and also brings you close up to some fascinating bedding and slumping in volcaniclastic rock outcrops. It is an ideal walk for those days when the bigger fells are in the mist. However, it should not be dismissed out of hand on clear days as that is when the views are at their very best.

Following the described route on the mountain summit area is fairly straightforward as a stone wall or fence accompanies you most of the way. Even if you stray off to look at features here and there it is easy to recover your position. However, if the summit area is in the mist, it is advisable to keep the wall/fence in sight at all times and not wander off to the sides, as the whole area is immensely complex. Off the summit ridge itself you are often faced with a tilted, confusing landscape of knobbly 'lochan and lochan' terrain with numerous dips and rises, and to make matters worse there are also some unguarded quarry faces.

Several generations of slate quarrying operations can be seen including present-day activities. Please take care in and around quarry workings. The described route takes you past some small-scale workings, now redundant, where some of the waste tips contain many fascinating small rocks with fine bedding revealing their sedimentary origins. Farther along the summit and on Side Pike, the walk crosses lava and pyroclastic rocks with more graphic rocks to be seen.

The return part of the walk follows paths and old trackways along the southern side of Great Langdale and provides excellent views of the glacial valley and surrounding mountains as well as the farmscape that has tamed the valley floor. Farming here has gone on for centuries and its mark is deeply etched into the landscape left by the ice age. However, the slate quarries that are passed on the start and towards the end of the walk remind us that the most dramatic changes to the landscape have occurred in the last two centuries.

Lingmoor Fell

From the bus stop, head past the YHA hostel. Bear right at the first junction. The tarmac track eventually becomes a stony way, rising steadily uphill to a gate. Go through the gate and continue along the track, the angle of ascent now easing off. Here you emerge from the trees and get views towards the eastern end of Wetherlam and the quarries on its lower flanks, over the hummocky terrain to your left and straight ahead and the rearing crags of Lingmoor Fell on the right.

Turn right through a gate at about 3177 0386 and follow the track which rises gently to the first outcrops at about 3150 0387, where you can see coarse tuffs. Go through a gate which comes into view and then rise more steeply up to a gap in the crags above and ahead. As you get higher you can see well-cleaved rocks as well as more coarse tuffs. On reaching the gap, the track bears left and carries on up. Over to the right you will see a stone wall which roughly marks your route all the way to the summit, although the track is often some way to the left of the wall and it may go out of sight now and again

A short way up, you should see a small quarry waste tip piled up next to the stone wall; have a mooch through the stones discarded here by the quarrying operations. You will need to pick up and turnover the lumps to see the different faces. Many of the rocks, where weathering has been limited, display incredibly beautiful bedding patterns, often curving and cutting through one another. These are off-cuts of the famous slates from the Seathwaite Tuffs, now known as the Seathwaite Fells formation (also previously known as the Tilberthwaite Tuffs). You can see large, polished versions in the quarry and showrooms at Skelwith Bridge between

Lingmoor Fell

Elterwater and Ambleside, also accessible by extending the last part of the walk alongside the river from Elterwater to Skelwith Bridge.

All the slate workings which you see on today's walk are exploiting rock from the Seathwaite Fells formation. It used to be believed that these were all tuffs formed from very fine-grained volcanic 'ash', but now geologists assess them to be partly 're-worked' material, i.e. tuffs that have been eroded (very shortly after being laid down while still fairly soft and thus easily eroded) then deposited in a lake which was created after the collapse of the Scafell caldera. Photo w13.1 shows an old slate quarry face exposing the cleavage plane; the slate rocks are heavily encrusted with lichen suggesting that the quarry has been disused for some time.

The Seathwaite Fells formation is one of the most widespread rock formations in the Upper Borrowdale Volcanic Group and was created towards the end of the volcanic cycle. The rocks of the formation include breccias and tuffs laid in water as well as reworked volcaniclastic sandstone, siltstone and mudstone. The water was thought to be a lake or lakes which formed in calderas and fault-bounded depressions which were created after major pyroclastic eruptions.

Follow the track up to the next high point (one of a series on the way to the summit) where there is a small shelter on the left made out of slate waste. This makes a good spot to stop for a break. It's worth looking at the slabs making up the shelter and spread around as you can see interesting bedding features such as graded bedding where larger lumps have fallen together to the bottom of the bed with smaller lumps above them. All these slates are part of the Seathwaite Fells formation even though these are quite different from the very fine-grained rocks that we saw at the last waste tip. Here the slates are more coarse-grained and more uniform in colour.

The views from this spot to Loughrigg Fell, Windermere and Wetherlam are excellent; note also the hummocky terrain in between. If you go to the front of the little work area overlooking the slope which you've just come up, you'll see that the terrain below consists of a hummocky terrace with plenty of quarry workings evident. Indeed, you may be able to hear the noise of the modern day quarry workings over to your left between here and Chapel Stile. Continuing up the track, over by a gap in the stone wall blocked by two iron gates, you can see the modern quarry workings way down below.

Scrabble up the rock outcrop ahead with great care as there are unguarded cliffs of an old quarry face on your right (Lingmoor Quarry at about 3085 0425). From here you begin to get views along the main ridge. Walk though the hummocky terrain with outcrops of cleaved rock. More old quarry workings can then be seen to the left in a wide dip. The line of the

Photo w13.2 | The summit plateau of Lingmoor Fell: an outcrop formed from an intrusive sill of basaltic andesite runs across the plateau in the centre of the photo above the scree slope.

Rosset Gill fault passes some way to the left, parallel to the general summit plateau, and is responsible for the shape of the fell farther left.

Looking again towards the summit beyond the wide dip with the quarry workings, in the rise on the other side of the dip at about 3060 0435, a quite different rock type is seen. This is a darker rock with no cleavage or bedding, which can be identified from the collection of scree at the base of the rise and exposed rocks at the top (see Photo w13.2).

This is a sill of basaltic andesite that has been exposed at the surface, having been intruded into the Seathwaite Fells formation rocks. Once at the top of the rise the rocks become part of the Seathwaite Fells formation again, with cleavage, bedding and coarse tuffs visible in the outcrops (see Photos w13.3 and w13.4). A short way on near the stone wall on the right is an outcrop standing upright on its own. This is well worth a visit as it shows bedding in the volcaniclastic rock, but also a widely-spaced set of cleavage planes cutting diagonally across the outcrop.

At around 3050 0425 the rocks change again, now becoming andesitic lavas. These are darkish and undistinguished rocks with no bedding or cleavage, but only cooling joints. The route accompanies these lavas for the next 850m.

My recommendation is to carry on past the rather undistinguished summit for another 200m to a point where the summit ridge drops down slightly. The views here (at about

Lingmoor Fell

Photo w13.3 + Photo w13.4 | Bedding and cross-bedding in sedimentary rocks of the Seathwaite Fells formation; cleavage runs from upper right to lower left.

3015 0465) towards the head of Langdale are truly magnificent, the scale being just perfect for gaining an overview of the mountains arrayed ahead of you from left to right. Here you can try to imagine glaciers pouring out of the valley heads or over the low points such as either side of Rosset Pike during the latter stages of the ice age (after the ice sheet had melted).

This viewpoint is surely one of the best in the area and certainly one requiring minimal effort to get here (the summit of Lingmoor Fell being only 469m high). This relatively low viewpoint is what contributes to the power of the scene laid out before you. This is a truly fine place to linger, soaking up the views of this enchanting landscape – weather permitting of course as it can be incredibly windy on this summit ridge.

Carry on until the next slight head at about 2985 0502 (just after a little dog-leg in the wall). This marks the end of the andesitic rock and the return of the Seathwaite Fells formation. Stick close to the wall for a while from here down as there are some really fascinating bedding patterns in the rocks, but also keep an eye out on the outcrops to the right as you drop down. Indeed, if you have the time and enthusiasm it pays to explore the outcrops off to the right for amazing sedimentary features.

At about 2973 0518 you start to meet brecciated rocks, some with quite big lumps as well as some smaller ones. There are also very good examples of bedding and slumping in outcrops to the right just on from here.

Shortly after, you will come to a short steep section. A few years back I was here on an extremely windy day and, just as I was about to start the short climb down, I was hit full on by a very strong gust of wind. This started me off on a continuous, body-juddering sneezing

Photo w13.5 | The Langdale Pikes from the Summit ridge of Lingmoor Fell: Side Pike (centre left).

fit that lasted for the next 10 minutes before I could continue on my way. It was a most strange experience, which I never had before or since and certainly I hope never to repeat it. It underlines that, despite its modest height, this is still a real mountain walk and one that

Photo w13.6 | Graded bedding in sedimentary tuffs of the Seathwaite Fells formation.

Photo w13.7 | Breccia and sandstone; the large and small lumps forming the breccia were erupted and fell down into sedimentary rocks.

Lingmoor Fell

can bring one face to face with dangerous weather and other hazards normally associated with higher peaks.

This steep section marks the transition from the Seathwaite Fell formation to the Lingmoor Fell formation, made up of 'dacitic' tuffs, coarse tuffs and breccias. After the steep descent you have to cross to the other side of the fence and then continue to descend to the

Photo w13.8

The route passes through this narrow gap on the flank of Side Pike. This is also a good example of how oversteepened cliffs peel away from the main rock face when glaciers melt away.

col before Side Pike. Here the rocks change again, but even more rapidly for Side Pike is part of an area (including Blake Rigg on Pike o'Blisco, see Walk 15) of wildly mixed rocks. It is the result of slumping of massive blocks (some up to 500m in length) of semi-solidified volcanic material on sloping ground on the base of the Scafell caldera, the slumping being sparked off by earthquakes and land movement along fault lines.

The way ahead has, I'm afraid, a width restriction as it is necessary to squeeze through a narrow gap (see Photo w13.8). Alternatively, descend and make your way back up farther round. This brings you out onto a narrow track that traverses round the south-western side of Side Pike to meet a track on its western, less steep face. This is the only way up to the summit of Side Pike for walkers and once up there it is necessary to turn around and come back down the same way, then to continue on down on the track towards the minor road below.

The summit of Side Pike is made up of andesite, then as you descend you cross bands of dacitic tuffs, basaltic andesite, more dacitic tuffs, welded tuffs, volcaniclastic sandstone and more welded tuffs. You can see brecciated lava (Photo w13.9) and pyroclastic breccia (Photo w13.10). This variety of rock is due to the western flank of Side Pike being part of an area where massive blocks of rock up to 500m in size slumped and slid down a slope within the Scafell Caldera. An intensely mixed range of rock types can be found in a small area – quite an unusual geological phenomenon (see Walk 15).

You may also spot some sedimentary rocks on the descent from the summit of Side Pike towards the col. A close look reveals small faults dissecting the beds. Sometimes you can also see the way that the beds have been displaced on either side of the fault; one side or the other has moved up or down a short distance. An example is seen in Photo 3.13.

Just before meeting the road, turn right onto the public footpath which heads down towards Great Langdale, with excellent views across the valley to the Langdale Pikes and the gills dug deeply into the fellsides. Turn right along a narrow track just before reaching a wood, eventually joining a more defined permissive path and then, from Side House, a public footpath. Somewhat later you pass a good example of an ice-polished rock. A track (starting at 3115 0495) allows an optional diversion to the remains of Spoutcrag Quarry which can be visited (with great care). Continuing along the main route, the track becomes a minor tarmac road from Bays Brown, cutting through woodland above a working quarry. Note the densely moss-covered rocks in the stone walls alongside the track. The road eventually joins the minor road just above the YHA hostel, and returns to the start point.

Walk #14 Wastwater

START ▶ TRACK TO WASDALE NATIONAL TRUST CAMPSITE (1805 0760)

FINISH ● CIRCULAR ROUTE

TIME ☕ 4 HOURS+

GRADE ◐ NAVIGATION ●

☁ TERRAIN ● ● ● ● ●

⊗ SEVERITY ● ●

When I was last at Wastwater researching for this book, the view along the lake from near Wasdale Hall towards the cluster of fells above Wasdale Head (Yewbarrow, Kirk Fell, Great Gable and the Scafell massif) was voted 'Britain's favourite view' by viewers of BBC television.

Although Wasdale Head is located in a difficult to access corner of Lakeland, it remains high on the wish list of every fellwalker given the wide range of hill routes on offer at the head of this sublime valley. This walk aims to draw attention to the lake itself as an additional attraction to fellwalkers, either on those inevitable cloudy days or on the day of arrival, after pitching your tent or booking in to your accommodation. The walk can be comfortably completed in four hours or thereabouts, but it can also easily be made longer by frequently taking time to sit and appreciate those magnificent views, or by spending time exploring the rocks alongside the route.

The first half of the walk is largely dominated by the presence of the lake a few metres below you on one side and a rearing mountainscape or slopes of scree on the other side. The Wastwater Screes are immensely impressive, if presenting some challenges to the walker. The scree itself is made up of lumps of lavas, but later the walk brings you close to granite and pyroclastic rocks. It is the overall feel of this immense glacial valley that is the key feature of this walk, however.

Most of the walk is pretty straightforward, presenting no real navigation problems and much of it on very easy terrain. However, I have given it the highest ranking for terrain as there is a 200m stretch of the walk across large boulders, part of the famous Wastwater Screes. This section requires care and balance. Careful selection of the route is also needed. None of this should prove to be any real problem to the experienced hillwalker, but may prove difficult for those not used to scrambling among rocks (and for dogs). A fall is unlikely to be fatal, but could well mean sustaining a leg injury in a very awkward place.

Turn onto the track leading to the campsite, cross the bridge and carry on along the track past the campsite entrance to cross another bridge 250m further on. Follow the track towards the lakeside and past Wasdale Head Hall Farm. After a while you reach a superb viewpoint (for looking back towards Wasdale Head) where the track rises above a small ridge (see Photo w14.1).

The track then passes through easy sections of scree. The rocks forming the hillside above are mainly lavas and intrusive equivalents, part of the Lower Borrowdale Volcanic Group eruptive sub-cycle when vast quantities of lavas poured out of volcanic vents. The screes are well-encrusted with lichen so it is usually difficult to identify much from the lumps of scree themselves, although they are generally clearly darkish rock. In places, it is possible to see fresh faces where a rock has fallen and broken open.

Photo w14.1 | Looking back towards Wasdale Head from shortly after the start of the walk:
Yewbarrow (left), Kirk Fell and Great Gable (centre) and Lingmell and lower flanks of Scafell Pike (right).

The screes are the product of freeze-thaw action in the post-glacial era, prising lumps from the rocks above which were left exposed when the glaciers melted. The screes mask the

Photo w14.2 |
The Wastwater
Screes looking
upwards from
near the lake.

fact that the hillsides are much steeper than they seem. Notice how the scree slope carries on below water level until it disappears from view. The bottom of the lake is lower than present day sea level. The glacier was at its most powerful here and was constricted by the tough rocks on either side, so it dug deep into the rock floor. Farther on, its power was lessened by the widening of the valley, so the glacier dug less deeply.

Eventually you reach the difficult section. The track disappears and it is necessary to pick your own route. There are some small cairns, but they are few and far between and easy to miss. The best route is probably at about 30-40m above the lake level, so avoid following the small track that descends towards the lake seen just as you approach the boulder field. Instead, go up about 10m before starting to cross the boulders.

The last time I was here I met a not very fit-looking couple struggling across the boulders accompanied by a large and decidedly unfit dog which they had to man (and woman) -handle across the great holes between the tumbled boulders. As we talked, one of them said that as it was called Wastwater Screes, they had expected scree not enormous great boulders. They had crossed about half of the section when I met them, so there was no point in their turning round. I kept an eye on them to ensure that they crossed the difficult section without mishap, but it must have taken them well over an hour to cross this short section of the route.

Even the fit and supple will be relieved when they can again put feet onto the ground, not least as you can again begin to attend to the wonderful views around and about rather than on where to place your feet and hands.

The route then takes you past a gorgeously quiet and tranquil spot where the lake de-bouches its water into its outlet stream, the River Irt. Keep on the track close to the river bank until it takes you to Lund Bridge at 1415 0390. Cross the bridge and follow the permissive path back along the other side of the river towards the lake again. After a couple of hundred metres the views open out along the lake. You will be especially lucky if the waters are tranquil and the fells are reflected in the lake, but even on misty days the views here are enchanting.

Continue along the track in wooded ground up to a wall and stile at 1475 0480. This brings you straight up to a rock outcrop.

This is a rock type we have met on only one other walk (on Walk 12 on Low Rigg) – granite. The next section of the route takes you past a series of granite outcrops for the next kilometre or so. The rock has a pinkish tinge to it that is quite easy to spot. Geologists believe that a vast mass of granite underlies much of the Lakeland district. This is based on gravity anomalies, which allow geologists to work out what rocks lie below the surface based on very precise gravity readings. This granite outcrops on the surface only in a few places, mainly in

Photo w14.3 | The Wastwater Screes.

the south-western fells. Where it does outcrop, it produces generally less impressive terrain than the other volcanic rocks of the area (see Chapter 4).

From here on, it is necessary in places to follow the tarmac road, but you can also leave the road in many places to walk alongside the shore of the lake or to cross the innumerable small outcrops. After a kilometre or so, a short distance before reaching Smithy Beck, the geology maps show that you cross into outrops of basaltic andesite. This is not immediately apparent however, as when the granite was intruded below the andesite it was of course still molten and hot. Not reaching the surface it cooled only very slowly. This had the effect of heating and toughening the surrounding rock, the basaltic andesite, in what is known as a 'metamorphic aureole'. The rocks clearly lose the pinkish tinge and become recognisably dark-grey rocks some 300m or so after the rock boundary shown on geology maps.

The fells to the west of the lake are largely made up of basaltic and andesitic lavas and intrusive equivalents with some outcrops of pyroclastic rocks, especially around the summit areas of Middle Fell and Yewbarrow. A big fault runs through Middle Fell, its line marked by Goat Gill, with pyroclastic rocks forming Goat Crag above and to the right of the gill.

The views to Yewbarrow's sharp ridge are superb. The pyroclastic rocks that make up the summit of Yewbarrow may represent the western limits of the Scafell caldera. (This may also

Wastwater

be along either a pyroclastic outcrop alongside Over Beck to the west of Yewbarrow or on the western flanks of Lingmell and Scafell.)

Carry on to just past Netherbeck Bridge, until shortly after the bridge there is a stile on the right. Cross the stile and follow the narrow track until it brings you to a quiet cove, looking directly at Brown Tongue (the ridge between two gills descending from below Scafell Pike). The right-hand side of the stream follows the line of a major fault, which continues up to form the col between Scafell Pike and Scafell, Mickledore.

It is possible here to see the rock boundaries on Illgill Head on the opposite side of the lake. The crags above the screes are all part of the Birker Fell formation, but the summit of Illgill Head is made up of pyroclastic rocks. This can be seen in the outcrops at the point where the slope becomes steep and the long ridge of Illgill Head starts to descend towards Wasdale Head Hall Farm to the left.

Resume the route either on the road or, with some difficulty, along the shore. The views of the mountains around Wasdale Head are superb, but increasingly one's attention is drawn to Yewbarrow's sharp southern aspect. As mentioned above, the summit area of Yewbarrow is made up of pyroclastic rocks, which now form a cap over the slightly less resistant lavas and intrusive equivalents. The final part of the route follows the tarmac road back to the start point.

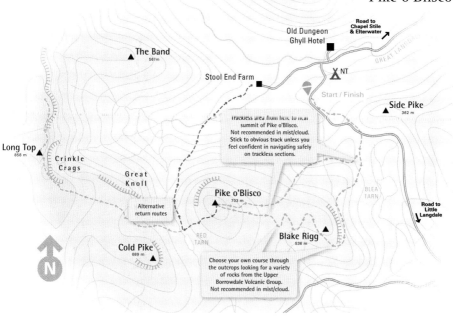

The Band
567m

Old Dungeon
Ghyll Hotel

Road to
Chapel Stile
& Elterwater

GREAT LANG

Stool End Farm

NT

Start / Finish

Side Pike
382 m

Long Top
858 m

Crinkle
Crags

Trackless area from here to near
summit of Pike o'Blisco.
Not recommended in mist/cloud.
Stick to obvious track unless you
feel confident in navigating safely
on trackless sections.

Great
Knoll

Pike o'Blisco
703 m

BLEA
TARN

Road to
Little
Langdale

Alternative
return routes

RED
TARN

Blake Rigg
536 m

N

Cold Pike
689 m

Choose your own course through
the outcrops looking for a variety
of rocks from the Upper
Borrowdale Volcanic Group.
Not recommended in mist/cloud.

Walk #15 Pike o'Blisco

START ▶ OLD DUNGEON GHYLL HOTEL, CAR PARK & BUS STOP, 285 059

FINISH ● CIRCULAR ROUTE

TIME ◔ 5 HOURS

GRADE ⊙ NAVIGATION ● ● ● ○

⌂ TERRAIN ● ● ● ○

⑤ SEVERITY ● ● ● ○

This walk is designed to introduce the fell-walker to a variety of pyroclastic rocks all within a single short walk on a stupendous fell. Various types of welded tuffs and breccias, 'accretionary lapilli' and gaseous cavities, and volcaniclastic sedimentary bedding and slumping can all be seen in two fairly limited areas. This makes it an excellent walk for getting to grips with a considerable variety of the more conspicuous of the different types of pyroclastic rock within the Upper Borrowdale Volcanic Group (BVG).

Pike o'Blisco

Pike o'Blisco's pointed summit is easily visible from below, within Great Langdale, and sits atop a sloping summit plateau area with distinct rock steps. Around its summit, its rocky ribs stick out like those of an underfed/healthy person. The ribs clearly tilt gently giving the summit area its overall feel. This makes it fairly straightforward to get an impression of the upper part of the mountain's geological structure.

The area around Blake Rigg, on the eastern flank of Pike o'Blisco, is particularly rich in different types of rock. It is part of a rather angular hourglass-shaped area with a narrow waist and wider top and bottom bounded by deep faults to the west and north. The northern-most fault forms the top and runs through the floor of Great Langdale near the start point of the walk. Blea Tarn and the western flank of Side Pike are included within the area, as is the south-eastern corner of Pike o'Blisco between Black Crag and Blea Moss. The summit area of Pike o'Blisco is outside this area.

The area is known as the Side Pike Complex and covers about five square kilometres. The whole area is made up of massive blocks of rock, some more than 500m long, which have obviously undergone some pretty violent movements. The blocks are badly deformed near their edges and are mixed up chaotically. There are signs that some big blocks have been driven into other blocks when the latter were still relatively soft (i.e. fairly soon after they were erupted and before having time to solidify to any great extent). These blocks show severe deformation as a result of the impact. The rocks have also been turned upside down in places.

This mish-mash of rocks is the product of violent eruptions, earthquakes and massive earth movements, causing big blocks of rock to break off from their position and to slump down a slope or newly created depression within a caldera. The blocks have then crashed and smashed into one another, creating the great treat of different rock types we see today.

The mapping of these different blocks is a job for professional geologists. However, the alert fellwalker gets the opportunity to see a considerable variety of pyroclastic rocks in an incredibly small area. An impression of the size and power of the volcanic eruptions, caldera-creation and volcano-tectonic faulting responsible for creating this geologically frenzied area can be gained fairly easily.

Pike o'Blisco is a highly attractive mountain in its own right and deserves to be climbed for the super views from the summit. It is a fairly compact mountain and can easily be climbed and descended in a short day's walking. Do remember, however, that taking time to look at the different rocks and to appreciate the surprising variety of rock types to be found within a small area can make this a much longer walk.

Most paths head straight for the summit and miss out the eastern flank, Blake Rigg. This is a pity as the many small crags around Blake Rigg (Bleaberry Knott, Long Crag and Wrynose Fell) are where the greatest variety of pyroclastic rocks is to be seen. This walk is designed to take in these usually unfrequented stretches as well as the rockier parts of the summit area to give a much better appreciation of this fine mountain.

The directions I give are somewhat vaguer than for most of the other walks as it would be impractical to give precise details to follow around and over innumerable rock outcrops. Instead I indicate a general direction to be followed, but leave it up to the reader to decide how many outcrops to look at. However, I can say from my own experience that the more outcrops that you look at, the greater the variety of rocks you will see.

There are no tracks in the Blake Rigg area and you will have to determine your own route. While it is possible to navigate your way around the outcrops of Blake Rigg in the mist using the OS 1:25,000 map, it would not be an easy task however experienced a map reader you are. This is probably not the best place to teach yourself these skills as there are steep crags to the north, the south and especially east of Blake Rigg. So, it's probably advisable to keep this walk for clear days, when the summit of Pike o'Blisco can be seen from most of the area and used as a reference point for navigating your way round as many outcrops as possible.

Continue west along the tarmac road from the bus stop, bearing sharp left after 100m. Continue until just after a hump bridge then turn left into the NT campsite on a public footpath. The footpath bears sharp right after another 100m then climbs through a succession of woods and fields until you come out into an open fellside. A highly conspicuous (and very slippery) path winds its way up towards the col between Side Pike and Pike o'Blisco.

On reaching the col, cross the stone wall on a large and awkward wooden stile and cross the tarmac road. Two tracks leave this col. One heads on a level course straight to the main track (which starts lower down the tarmac road). A second higher track can be seen heading up where it joins the main track out of sight. The recommended route initially heads up towards this second track, but bears left shortly after reaching it.

Pike o'Blisco

Head uphill, keeping a stone wall a short distance to your left then follow the vague higher track for its first section only. Where the track peters out, strike left up steep grass between outcrops towards a gully high up on the left with what looks like some fearful crags.

Keep an eye on the rock outcrops as you ascend as there are some striking breccias (see Photo w15.1) and also sedimentary features to be seen. The gully will eventually bring you out near the letters 'ke' of Blake on the OS 1:25,000 map. The fearful crags look less impressive and are just one more outcrop of many all around you now.

From here you can pick your own route, visiting as many or as few of the outcrops and minor prominences as you wish. You could head slightly away from the direction of the summit to look at the outcrops at the top of the crags of Blea Rigg overlooking Little Langdale.

Pick your way, more or less direct, through the crags and outcrops of Blake Rigg, Bleaberry Rigg and Long Crag, heading generally towards the summit of Pike o'Blisco. It goes out of view behind crags at times, but only ever temporarily.

The more outcrops you peer at, the more rock types you will see (see Photos w15.2–w15.7). The 'grain' of the outcrops is such that many of the most interesting rock faces are pointed in the direction in which you are heading, so it is always worth looking back round at each outcrop as you pass round/over/through it. The rock outcrops provide the opportunity for some easy scrambling if desired. You can see examples of breccia, coarse tuffs, bedding, slumped bedding, welded tuffs and 'accretionary lapilli' (formed by the accretion of material immediately after the eruption which then settles into sedimentary deposits). These represent about as many of the different types of Upper BVG pyroclastic rocks that you will see on all the other walks combined.

Photo w15.1

Outcrop (about 3m high) with breccia at the base and bedding at the top, below Blake Rigg.

All photos on this page were taken between Blake Rigg and Pike o'Blisco summit.

Photo w15.2 | Breccia.

Photo w15.3 | Slumped bedding.

Photo w15.4 | Wider view of bedding, including slumped bedding.

Photo w15.5 | 'Accretionary lapilli' formed by material accreting after the eruption then settling in sedimentary tuffs.

Photo w15.6 | Welded tuff.

Photo w15.7 | Tuffs formed by settlement of pyroclastic fragments in water; the different beds are clearly visible.

Pike o'Blisco

The area around here is a good example of small-scale hummocky terrain where glaciers have plucked out rock at the boundaries between different layers. There are plenty of small tarns nestled in the little dips between outcrops, adding to the interest and beauty of the walk (see Photo w15.8) and plenty of vistas onto nearby mountain ranges (see Photo w15.9).

Photo w15.8 Small glacial tarn on Long Crag.

Photo w15.9 View of Swirl How and Wetherlam from Long Crag.

After passing Long Crag and approaching the next rise (where it says 'Wrynose Fell' on the OS 1:25,000 map) you should be able to see the main track coming in from the right. Higher up in the direction of the summit, you should also see an area of flat, tilted rock exposed at the surface, a bedding plane of sedimentary rock. Head towards this outcrop and then move right to join the main track for the final part of the ascent to the summit. (It is possible to head straight up towards the summit area over grassy terrain, but joining the main track takes you past more interesting rock outcrops.)

Shortly after joining the main track you meet a ten metre high rock face which requires a short scramble. Note the bedding in the rock just before you start to scramble. Another rock

Pike o'Blisco

Photo w15.10
View towards the summit from near the main track, showing layers of tilted rock that have to be crossed.

Photo w15.11
View from the summit of Pike o'Blisco (looking south-east), showing the summit layer of lava.

face and short scramble follow shortly after. These are some of the tilted rock beds that you can see from the valley below. From the second scramble, the going is fairly easy. The summit views are superb and there are plenty of places where you can sit out of the wind if conditions are not ideal. It's well worth taking a good time to wander around the summit area and enjoy the range of views, weather permitting of course (see Photo w15.11).

You may have seen a wide variety of pyroclastic rocks on the way up but the actual summit area of Pike o'Blisco is made up of lava, as are the slopes to the west leading down to Red Tarn.

You can either descend by following the main track back towards Great Langdale or, for a marginally longer route, descend towards Red Tarn and then via Oxendale and Stool End farm back to the beginning. If you want to make a very long day of it, then carry on over Crinkle Crags (plenty of examples of welded tuffs and superlative views into Eskdale) to Three Tarns, then descend via the Band.

Acknowledgements

Thanks go to Michael Evans, Emma Williams, Paul Davis, Antony Ryan, Charlie Marshall and Clair Drew for reading and commenting on parts of the text of this book and for making many useful comments. Reg Atherton acted as a 'guinea pig' for some of the walks providing valuable reader reaction. Thanks to Ian Smith, Reg Atherton and Jim Young for providing photos of glaciers and volcanoes from distant parts, to Ian Tyler of the Keswick Mining Museum, to Peter Wood and Franco Ferrero of Pesda Press for enthusiastic support of the Rock Trails books, and also to Don Williams at Bute Cartographics. In Llanberis, thanks also to Sam and staff at Saffron, Al and Rachel at Georgio's, Stefffan at Pen Ceunant café, Tim at High Ropes, Dylan at Crib Goch, Peter at White Spider, Jim Langley at Nature's Work, and to Phill and Lisa George.

Photo copyright: All photos by the author except Photos 2.2 and 5.4 by Jim Young, Photo 6.1 by Ian Smith, Photo 5.3 by John Simmons, the Geological Society of London, and Photo 7.4 by Reg Atherton.

Geology map copyright: Reproduced by permission of the Yorkshire Geological Society and the British Geological Survey. Sketch maps in this book are based on the — Ordnance Survey 1948 Lake District map.

Pattern swatches based on Public Review Draft Draft Digital Cartographic Standards for Geologic Map Symbolization (PostScript) by the US Geological Survey (Open File Report 99-430).

Bibliography

J Adams, *Mines of the Lake District Fells*, 1995.

M Atherden, *Upland Britain: A Natural History*, 1992.

D Cannadine, *G M Treveleyan: A Life in History*, 1992.

H Davies, *Wainwright: The Biography*, 1995.

M Dodd (ed), *Lakeland Rocks and Landscape*, 1992.

M B Donald, Elizabethan *Copper: The History of the Company of Mines Royal 1568–1605*, 1955.

E G Holland, *Coniston Copper Mines: A Field Guide*, 1981.

A Hunter and G Easterbrook, *The Geological History of the British Isles*, 2004.

B Lynas, *Lakeland Rocky Rambles: Geology Beneath Your Feet*, 1994.

D Millward and A Robinson, *The Lake District*, 1970.

D Millward and A Robinson, *Upland Britain*, 1980.

D Millward et al., *Geology of the Ambleside District*, BGS, 2000.

G H Mitchell, *The Lake District*, 1970.

F Moseley (ed), *The Geology of the Lake District*, 1978.

F Moseley (ed), *The Lake District*, 1990.

W A Poucher, *The Lakeland Peaks*, 1960.

R Prosser, *Geology Explained in the Lake District*, 1977.

E H Shackleton, *Lakeland Geology*, 1981.

M Shoard, *A Right to Roam*, 1999.

R Solnit, *Wanderlust: A History of Walking*, 2001.

I Tyler, *The Lakes and Cumbria Mines Guide*, 2006.

A Wainwright, *A Pictorial Guide to the Lakeland Fells: Book 1*, 1955.

A Wainwright, *A Pictorial Guide to the Lakeland Fells: Book 4*, 1960.

A Wainwright, *A Pictorial Guide to the Lakeland Fells: Book 6*, 1964.

N Woodcock and R Strachan (eds), *Geological History of Britain and Ireland*, 2000.

D G Woodhall, *Geology of the Keswick District*, BGS, 2000.

Glossary of Geological & Geomorphological Terms

Breccia – volcanic rock containing visible lumps of rock.

Caldera – roughly circular-shaped depression caused by volcanic explosion

Continental crust – a section of the earth's outer surface, the crust, that forms a land mass. The edge of a continental crust usually extends some distance out to sea, with a continental shelf. About 25% of continental crust is below current sea level as the boundary with oceanic crusts is fairly deep undersea. Continental crusts are less dense and more buoyant than oceanic crusts, but also thicker – the average thickness of continental crusts is about 35km and can be up to 75km thick where new mountain ranges are being built (such as the Himalayas). The crust rests on top of the earth's outer mantle and together with it forms the 'lithosphere' which is split into 'tectonic plates'.

Erosion – the physical breakdown of rock by water and ice. The rate of weathering (i.e. chemical breakdown of rock) and erosion is often related to climate (low temperatures lead to glaciation, hot ones to desertification, for example, each with their own patterns of erosion and weathering), but the relationship is complex and controversial.

Fault – a crack in the earth's surface, ranging from quite small to very large, where rocks move in relation to one another. Faults also provide routes to the surface for molten magma lying under the surface.

Fault line – the surface expression of a fault; also giving terms such as fault-line scarp and fault-line lake.

Geomorphology – the scientific study of the form of the land surface and the processes which affect it at the interface between the lithosphere, the atmosphere, the hydrosphere and the biosphere. While geology is largely concerned with the creation of the underlying rocks, geomorphology concentrates on the forces that shaped the rocks into the landscape.

Glacier ice – the dense 'blue ice' of which glaciers are made up has an air content of less than 20%; firn contains 20-30% air, 'granular ice' (a pre-firn stage of ice development) 30-85%, and snow 85-90%. As snow accumulates over the years, due to its not melting during cold summers, the snow particles lower down are compressed and deformed, subject even to some partial melting and refreezing into ice made up of smaller particles. These particles coalesce to form firn. Continued snow accumulation leads to further compression and the creation of solid and dense glacial ice.

Igneous or volcanic activity – falls into two main types: first, 'extrusive' where magma erupts on to the surface of the earth (on land or under sea level) either as molten lava or as explosive 'pyroclastic' fragments; and second, 'intrusive' where the magma doesn't reach the surface but is intruded between layers of soft rock below the surface (and may later be exposed at the surface by erosion).

Igneous rock – see 'Rock types'

Ignimbrite – rock produced by highly explosive and extremely hot pyroclastic eruption; also known as 'welded tuff'.

Isostatic rebound – during glaciation, the weight of the ice compresses the crust and upper mantle. After the ice has melted the land surface rebounds. The same effect occurs, but on a much longer timescale, as rocks are eroded and weathered away. Isostatic rebound from the weight of ice during the ice age is taking place today, so that the mountains of Lakeland are rising just about as much as they are eroded away.

Lithosphere – the earth's crust and the upper part of the mantle (lithosphere means the 'sphere of rocks'). The lithosphere is divided into a set of interlocking 'tectonic plates'.

Magma – molten (or partially molten) rock from the mantle, which accumulates in reservoirs under the crust when melting is caused by factors such as subducting oceanic plates. Magmas are chemically very complex and their nature depends on conditions; indeed they undergo change from the time of production to that of eruption. Non-viscous magmas, such as basaltic magmas, often produce lavas. Magma contains gases as well as elements. For example, gases are released as a result of chemical processes and are held in bubbles within the magma under pressure, waiting to escape as if opening a bottle of 'fizzy' water. Although often present in only small quantities, gases can have an important effect on the type of eruption if the gases cannot escape. Rhyolitic magmas tend to be quite viscous and this has the effect of retarding the escape of gases within the magma. As the magma rises up towards the Earth's surface, pressure reduces. More gases are released as water, carbon dioxide and other gases are formed in the liquid magma, creating more and more bubbles and/or increasing the size of existing bubbles. The viscous magma acts to retard the growth of the bubbles, increasing pressure within them. When it becomes too high compared with that of the encasing liquid magma, the magma is transformed from a liquid holding bubbles to a gas holding molten fragments (pyroclasts) of bubbly liquid (a process known as 'exsolution') which then creates the highly explosive eruptions of gas and fragments common with rhyolitic magma. Magma reservoirs may produce both rhyolitic and basaltic eruptions from the same magma source.

Metamorphic rocks – see 'Rock types'

Nueé Ardente – the type of eruption which produces welded tuffs and ignimbrites.

Oceanic crust – basaltic lava erupting at the surface, along with lower-lying intruded basaltic rocks (including 'gabbro'). New oceanic crust is formed which spreads outwards from the zone of eruptions (the 'mid-ocean ridge') to be recycled by subduction, where the oceanic crust collides with the rocks of other oceanic or continental crusts. Oceanic crusts are generally relatively young rocks (up to about 200 million years old at the margins) and are much thinner than continental crusts, but also denser.

Orogeny – the process of building mountain ranges caused by the collision of two continental plates. The plates buckle, crack and fold, building upwards to create mountains.

Outcrop – strictly speaking the bedrock or rock nearest the surface, but which may be covered by 'superficial' deposits (unconsolidated peat, alluvium and so on). However, I have used it to mean what it says, 'rock that crops out' visibly on the surface and through any superficial deposits. Geologists also use the term 'exposure' for this.

Plate tectonics – the theory that the Earth's surface is formed of independent but interlocking plates which are driven in circulation by convection currents within the underlying mantle (although there is no agreement on whether this is the sole, or even the major, cause). There are about a dozen major plates and several smaller ones, each moving independently of the others, sometimes colliding with each other. Plates can carry oceanic crust, continental crust or both. The plate boundaries are the site of most (but not all) volcanic activity (and also of earthquakes). It is currently believed that the continents join up into a supercontinent and then break up into continents cyclically over periods of about 500 million years. Events such as the opening and closing of the Iapetus Ocean and subsequent collision of Avalonia and Laurentia were part of such a cycle. At the time of these events the area was not in its present position, but was south of the equator, and has over the last 400 million years floated on the convection currents to the present locations. There were more minor earth movements too – the Lake District is further away from Snowdonia today than it was in Ordovician times (due to movement along major north-east to south-west trending fault zones). Another point worth noting is that collision is more likely to be oblique than fully direct. Plate movements are not all the same – the most rapidly moving plate, the Pacific oceanic plate, is currently moving at about 10cm per year, whereas the African continental plate is static.

Pyroclast – any fragment expelled in a volcanic eruption (as opposed to flow of molten lava) although the fragments can contain lumps of molten magma. The fragments can be very small (known as 'ash') to extremely large (several metres in size) and can be solid, liquid or gaseous (or a mix) at the time of eruption.

Rhyolite/rhyolitic – rhyolitic rocks have less iron and magnesium than basaltic rocks and more silica and potassium (rhyolitic rocks have over 66% silica, intermediate rocks between 52 and 66% and basic rocks less than 52%). This is believed to happen because iron and magnesium have a higher melting point than other elements, especially silica and potassium. So, when the mantle rocks start to melt, a rock of rhyolitic composition can be produced at lower temperatures than those of basaltic composition – if the molten portion can be separated. Rhyolite and rhyolitic are essentially synonyms, but do have different uses. Rhyolitic is applied to any 'acid' or rhyolitic rocks (rhyolitic lava, rhyolitic tuffs etc), whereas rhyolite is a name given to intrusive rhyolitic rocks. Rhyolitic rocks are more viscous, so flow less easily and cool down in vents. This leads to subsequent build up of gases and underlying molten magma, resulting in highly explosive eruptions. Rhyolitic rocks often weather to a light grey or even almost white aspect.

Rock types – rocks are usually divided into three great types though there is some overlap between the categories: igneous (or volcanic), sedimentary and metamorphic. All rocks start as igneous material, erupted or intruded by one means or another to form rock. Igneous rocks are very complex with over 1,500 different classifications, often overlapping and contradictory. Essentially igneous rocks can be 'acid' (i.e. rhyolitic or granitic), intermediate (andesitic) or 'basic'

(basaltic or doleritic). Oceanic crust is made up of basaltic rocks, while continental crust is usually made of andesitic or intermediate rocks (named after the Andes where they occur frequently). Rhyolitic rocks are comparatively rare, but do occure widely in central Lakeland. Roughly, acid rocks are light-coloured (due to a high proportion of light-coloured felsic minerals and are quartz and feldspar rich), while basic rocks are generally darker (due to being iron and magnesium rich). Currently, about 4 million cubic km of extrusive igneous rock and between 22 and 29.5 million cubic km of intrusive igneous rock are produced annually across the globe (from about 25 to 35 million cubic km of magma). The rocks produced by igneous activity can then be eroded and weathered into smaller pieces which are carried away, usually by water or ice, and deposited. They eventually form sedimentary rock, which can also be formed by wind-borne sediments of biogenic, biochemical and organic origin, such as limestone, and sedimentation of other materials, such as 'evaporites'. 'Detrital' grains are held in a much finer-grained 'matrix' and cemented by minerals that develop as a result of chemical processes. The processes of erosion, sedimentation and transformation into rock ('lithification') can be repeated several times. Nearly all the naturally occurring minerals can be rock grains, but the distribution is limited because some minerals and rock types are more stable, quartz being the most stable and thus the most common grain in 'sandstones'. Mudrocks are the most common of sedimentary rocks and are a mixture of 'clay' (grains less than 4 microns in size) and 'silt' (grains of between 4 and 62 microns). Igneous and sedimentary rocks can also be subject to high temperatures and/or pressures by later tectonic activity. This can cause changes to the minerals within a rock type, transforming it into a different type. Two important effects of metamorphosis are noted in sedimentary rocks in the Lakeland area: the transformation by pressure generated by mountain-building activity of mudstone into slate and the hardening of the rocks forming peaks or high areas by heating caused by intrusion of hot igneous rocks (forming a 'metamorphic aureole' in the surrounding rocks). Just as some igneous rocks are also sedimentary (for example, a fragmental/pyroclastic eruption will throw ash into the air which then settles on water and then drops down to form sedimentary rock), so too can the igneous/metamorphic label become a mite confused. For example, granite is continental crust rock which melts deep down in the lithosphere (due to pressure and heat generated by plate collision and mountain building) to form magma. However, it may not move upwards, so is viewed by different geologists as being either metamorphic or igenous. –

Tectonic plate – strictly speaking tectonic plates consist of the crust (either or both continental and oceanic) and the upper mantle (the lithosphere). For simplicity I have referred in the text to 'continental plate' and 'oceanic plate'.

Tuff – unfortunately the term 'tuff' suffers badly from being used very inconsistently by geologists, so I have simplified its use. Strictly speaking a tuff is the rock produced from volcanic ash. A volcanic ash is any pyroclastic fragment less than 2mm in size (it is not ash in the everyday sense of the term, the residue of combustion). Fragments between 2 and 64mm in size are called 'lapilli' and the hardened rock with such fragments is known as 'lapillistone'. Rock with fragments in excess of 64mm in size is 'breccia'. However, in recent years the intermediate lapilli/lapillistone terms have fallen out of fashion and most geologists now only distinguish between tuffs and breccias, but without updating the strict definitions. I have adopted this practice and only refer to tuffs, coarse tuffs and breccias, using a rough and ready means to distinguish between them i.e. any pyroclastic or intrusive rock with large visible lumps (from say the size of a grape upwards) is a breccia and any with fragments too small to be seen with the naked eye is a tuff, and anything in between (the size of pimple to the size of a grape) is a coarse tuff. There are

many types of tuff (and breccias too). There are 'air-fall tuffs', 'ash-flow tuffs' and 'ash-surge tuffs' and so on. A tuff can also have a small mixture of lump sizes and still be a tuff or tuffite (depending on the proportion of lapilli or non-igneous matter to be included in the mix). More confusion is added when welded tuffs are included, for they contain fragments ('lapilli'), usually of pumice, that have been flattened and welded into the smaller fragments, regardless of the size of the pumice fragments. To add yet further confusion, tuff is used for 'formation' names – the Seathwaite Tuffs (now the Seathwaite formation) - which include tuffs. Geologists today refer to lapillistone, breccias, siliceous nodule rocks, contemporaneously-intruded-rhyolites and sedimentary 'volcaniclastic' sandstone. The reader will no doubt forgive me for having simplified the use of tuff to mean all these things without great distinction, i.e. rock formed out of igneous fragments (pyroclasts). Breccia is identified in this sense as a sub-form of tuff. Purists may object, but in actual fact this is how the term is often used by geologists, despite the traditional formal definition. The definition I have used comes from the book *Volcanoes* by M Rosi and others (2003), 'deposited pyroclastic rock that has consolidated'. This does not include a definition of the size of the fragments, so I apply it to where the fragments cannot be easily distinguished by the human eye in rock outcrops.

Volcaniclastic sandstone – used to describe two types of rock (often indistinguishable even to the geologist in the laboratory. The first is volcanic ash which is erupted into the air and then settles on water, then drops down to form sedimentary rock. The second is volcanic rock which is erupted onto land and is then subject to erosion with the eroded fragments being carried away by rivers to be deposited at sea where sedimentary rock is formed.

Index of Place Names

Note: The grid references in the area covered by this book are in two grid areas NY (northern part) and SD (southern part). The dividing line runs just north of Harter Fell, south of Swirl How and Wetherlam, and north of the town of Windemere.

Gasgale Head